Computability Theory and
Foundations of Mathematics

Computability Theory and Foundations of Mathematics

Proceedings of the 9th International Conference on Computability Theory
and Foundations of Mathematics

Wuhan, China,   21–27 March 2019

Editors

NingNing Peng
Wuhan University of Technology, China

Kazuyuki Tanaka
Tohoku University, Japan

Yue Yang
National University of Singapore, Singapore

Guohua Wu
Nanyang Technological University, Singapore

Liang Yu
Nanjing University, China

World Scientific

EW JERSEY · LONDON · SINGAPORE · BEIJING · SHANGHAI · HONG KONG · TAIPEI · CHENNAI · TOKYO

Published by

World Scientific Publishing Co. Pte. Ltd.

5 Toh Tuck Link, Singapore 596224

USA office: 27 Warren Street, Suite 401-402, Hackensack, NJ 07601

UK office: 57 Shelton Street, Covent Garden, London WC2H 9HE

British Library Cataloguing-in-Publication Data
A catalogue record for this book is available from the British Library.

COMPUTABILITY THEORY AND FOUNDATIONS OF MATHEMATICS
Proceedings of the 9th International Conference on Computability Theory and Foundations of Mathematics

ISBN 978-981-125-928-9 (hardcover)
ISBN 978-981-125-929-6 (ebook for institutions)
ISBN 978-981-125-930-2 (ebook for individuals)

For any available supplementary material, please visit
https://www.worldscientific.com/worldscibooks/10.1142/12917#t=suppl

Printed in Singapore

Dedicated to Professor Chitat Chong
on the occasion of
his 70th birthday

Preface

The aim of Computability Theory and Foundations of Mathematics (CTFM) is to develop computability theory and logical foundations of Mathematics. The scope involves the following topics: Computability Theory, Reverse Mathematics, Nonstandard Analysis, Proof Theory, Constructive Mathematics, Theory of Randomness and Computational Complexity Theory.

The CTFM series has been headed by Kazuyuki Tanaka (Tohoku University). And CTFM has been held as the "Workshop on Proof Theory and Computability Theory" and have taken place in Matsushima (2008, 2009) and Inawashiro (2010). The previous meetings of CTFM took place in Sendai (2011), Tokyo (2012 to 2016), Singapore (2017) and Tokyo (2018).

The 9th Computability Theory and Foundations of Mathematics was held in Wuhan, China, from March 21–27, 2019. CTFM 2019 was organized and mainly sponsored by the Department of Mathematics at Wuhan University of Technology. The program consists of 1 Plenary Lecture by Prof. Chitat Chong (National University of Singapore), 16 Invited Talks and 10 Contributed Talks.

The editors wish to thank Wuhan University of Technology for giving us the opportunity to organize this event. We are grateful that the staff members at the Department of Mathematics in Wuhan University of Technology were so helpful and talented in organizing the venue, and taking care of everything for the smooth progress of the event. We are grateful to the members of the local committee, especially to Prof. Liang Zhang and Prof. Huansong Zhou for their hospitality and support. Our deepest gratitude goes to the volunteers whose dedication created a great atmosphere.

We are very pleased to find that *World Scientific* were enthusiastic to publish the CTFM 2019 proceedings. We appreciate all participants very much for their contributions in the workshop and to this proceedings. We acknowledge also the scientific reviewers who read the articles carefully and gave many important suggestions to the authors. We would like to thank the publishing editor Ms. Lai Fun Kwong and the designer for their help.

This book was partially supported by the postgraduate textbooks and monographs funded construction projects of Wuhan University of Technology and NSFC grant 11701438.

NingNing Peng

Opening Speech

Welcome to the 9th annual meeting of Computability Theory and Foundations of Mathematics. My name is Kazuyuki Tanaka. I am the co-chair of the organizing committee and the founder of CTFM. I really feel honored to have this conference in China for the first time. So, first of all, I would like to express my deep gratitude for the President of Wuhan University of Technology and all the people working for this conference.

Tohoku University is the third imperial university in Japan, but it is also well known as the first to admit foreign students such as Lu Xun and Su Buqing, and female students since 1913. The university is located in Sendai, the largest city in Tohoku, namely the northeastern district. Probably most of you know that a huge earthquake hit eastern Japan in March 2011. Actually, the coastlines were seriously damaged by Tsunami, but fortunately the Sendai city were almost safe.

The first meeting of CTFM was held in February 2011 in Sendai, just a few weeks before the earthquake. In addition, I had organized several predecessor meetings in the suburbs of Sendai before CTFM. Particularly around 2008, when I was chairman of Mathematical Institute, our logic group got pretty large by admitting many graduate students including NingNing-san and several postdocs, which even made our regular seminars almost like small international workshops.

After the earthquake, I held the second to the sixth meetings of CTFM in Tokyo, and thereby attracted more and more participants. Then, the seventh were in Singapore, and the eighth in Tokyo, again. Now, I am happy to have this 9th meeting in China with many of my former students and old friends. I hope this workshop provides everyone here with the opportunity to strengthen international cooperation and advance the new research of mathematical logic.

Finally, I would like to take this opportunity to express my deep gratitude to Professor C.T. Chong for his continuous assistance and friendship to me and CTFM. Although it may be too early, we would like to say Happy 70th Birthday to C.T. If I remember correctly, we met each other for the first time in PennState in 1989 or 1990. When we had dinner together with Steve Simpson, I asked C.T. about the past Asian Logic Conferences, which was mainly launched by C.T. in 1981. In 1990, Japan was preparing for its 4th meeting in the charge of Prof. Hirose, a recursion theorist close to me. But very sadly, he passed away before the meeting, which loosened my relation to the 4th ALC. However, at its 5th meeting in Singapore again in 1993, I became an invited speaker, and gave a talk on non-standard methods in second order arithmetic. There are many stories after this. But jumping to 2011, soon after the earthquake, six graduate students of mine were invited to a summer school organized by C.T., which really encouraged and strengthened our group so much. CTFM and my logic group had never been so successful without his help.

I will end my story here. Thank you very much for your attention. Now, I will hand it over to Chair NingNing. Thank you!

Kazuyuki Tanaka

Organizing Committee

NingNing Peng	Wuhan University of Technology, China, Chair
Kazuyuki Tanaka	Tohoku University, Japan, Co-chair
Yue Yang	National University of Singapore, Singapore
Guohua Wu	Nanyang Technological University, Singapore
Liang Yu	Nanjing University, China

Program Committee

NingNing Peng	Wuhan University of Technology, China
Kazuyuki Tanaka	Tohoku University, Japan
Ulrich Kohlenbach	Technische Universitt Darmstadt, Germany
Stephen Simpson	Vanderbilt University, USA
Toshio Suzuki	Tokyo Metropolitan University, Japan
Liang Yu	Nanjing University, China
Yue Yang	National University of Singapore, Singapore
Guohua Wu	Nanyang Technological University, Singapore

Contents

Overview of CTFM 2019

Chi-Tat Chong's Work on Reverse Mathematics
by Yue Yang

This desirable article reports on the illustrious research history of Professor C.T. Chong, who is an essential figure for the CTFM to be developed successfully. It also gives a neat overview of the fusion area of recursion theory and reverse mathematics.

On One-Variable Fragments of Modal μ-Calculus
by Leonardo Pacheco, Wenjuan Li and Kazuyuki Tanaka

In the paper, the authors study one-variable fragments of modal μ-calculus and their relations to weak parity games. They introduce one-variable transfinite formulas to characterize the Δ_2^μ formulas.

Infinite Games, Inductive Definitions and Transfinite Recursion
by Kazuyuki Tanaka and Keisuke Yoshii

The paper investigates the logical strength of determinacy of Gale-Stewart games in the difference class of Σ_2^0 sets. An axiom of transfinite recursion of inductive definitions is introduced to pin down the strength of some ambiguous classes.

A Survey of the Distributional Complexity for AND-OR Trees
by Weiguang Peng

This paper summarizes some recent results on the distributional complexity for multi-branching trees with respect to different kinds of distributions and classes of algorithms, especially the study of the eigen-distribution on ID for multi-branching weighted trees.

*Rational Sequences Converging to Left-c.e. Reals of Positive
Effective Hausdorff Dimension*
  by Hiroyuki Imai, Masahiro Kumabe, Kenshi Miyabe,
  Yuki Mizusawa and Toshio Suzuki

  The paper investigates effective dimensions and ideals closely
  related to quasi Solovay reducibility by means of the rate of
  convergence. It also gives a variation of the first incompleteness
  theorem based on Solovay reducibility.

Takeuti-Yasumoto Forcing Revisited
  by Satoru Kuroda

  The author develops the forcing method on nonstandard models
  of bounded arithmetic, which is initiated by G. Takeuti and M.
  Yasumoto. He also relates separation problems of complexity
  classes to properties of generic extensions.

Permission and Obligation in Ceteris Paribus
  by Huimin Dong

  The paper argues that permission and obligation should be
  viewed as the sufficient and necessary conditions for the Right,
  but should be governed under the principle of *ceteris paribus*.
  This viewpoint turns to a general formal theory of the dynamic
  deontic logic for permission and obligation, which can answer
  the Lewis problem, the Ross paradox, and others.

Reverse Mathematics of Separation Theorems in Lattice Theory
  by Junren Ru and Guohua Wu

  In the paper, the authors study reverse mathematics of countable
  lattice theory. They prove that the existence of $\mathcal{J}(L)$, the set of
  irreducible elements, of a countable lattice L, is equivalent to
  ACA_0, and also that **DPI**, prime ideal separation theorem, is
  provable in WKL_0.

Chi-Tat Chong's Work on Reverse Mathematics

Yue Yang

Department of Mathematics,
National University of Singapore, Singapore
E-mail: matyangy@nus.edu.sg

The 9th International Conference on Computability Theory and Foundations of Mathematics (CTFM) was held from 21 to 27 March 2019 at Wuhan University of Technology, Wuhan, China. The organizers had asked me to write an article about the mathematical research of Professor Chi-Tat Chong, on the occasion of his 70th birthday, because of his invaluable support for the CTFM series. This request gives me both pleasure and anxiety. On the one hand, I am very happy to take this opportunity to send my "birthday gift" to CT, as he is colloquially known by his colleagues and friends; on the other hand, it is impossible for me to write any meaningful comments on his research that would even satisfy the minimum standard. The reasons are manifold. First of all, CT is still working tirelessly and is steadily producing high quality research output. Secondly, his research interests are extremely wide; many of his works are in areas that I am not familiar with. Therefore, what I will do is to write what I know about his work on reverse mathematics; and state some facts about his other contributions. It is definitely subjective and perhaps only reflects a tiny fragment of the complete picture.

Chi-Tat Chong was born in Hong Kong on 6 September, 1949. His family moved to Singapore in 1958 after spending a few years in Indonesia. He obtained his Ph.D from Yale University in 1973 under the supervision of Manuel Lerman. The title of his thesis was: "Tame Σ_2 functions in α-recursion theory". He returned to Singapore in 1973 and began his career in National University of Singapore (NUS), which was then known as the University of Singapore, as a Lecturer in 1974. He became full professor in 1989 and University Professor in 2004. He was elected to fellow of Singapore National Academy of Science in 2011 for "his significant contributions to mathematical logic and the general understanding of computability".

Beginning from the 1980's, under the strong support of the Singapore government, NUS underwent a transformation from a local institution to a global university known for its research leadership. CT certainly contributed tremendously in this transformation, as he was Deputy Vice Chancellor, Deputy President and Provost of NUS from 1996 to 2004. Besides his contribution to the university, he has been influential in the mathematical community in Singapore.

CT is also played an important role in the logic community in Asia. He organized the first Southeast Asian Conference in Mathematical Logic at NUS in 1981, which today is recognized to be the inaugural meeting of the Asian Logic Conference series. He also set up the Graduate Summer School in Logic at Institute for Mathematical Sciences (IMS) in 2006, whose objective was to "bridge the gap between a general graduate education in mathematical logic and the specific preparation necessary to do research on problems of current interest in the subject". While a majority of the participants will be graduate students, the summer school also attracts postdoctoral scholars and researchers. Almost all young logicians in China had attended IMS summer school, and so had many young Japanese logicians. It is because of CT's long lasting influence in the Asian logic community that we celebrate his 70th birthday in this CTFM.

1. An Incomplete List of CT's Work in Recursion Theory

Let me list some of CT's research work in Recursion Theory that I know of. The topic of reverse mathematics will be discussed in the next section.

CT's main interest is in recursion theory, which traditionally focuses on computability of natural numbers. However CT investigates computability far beyond the traditional domain, his research topics cover all aspects of computation: From classical Turing degree theory to higher recursion theory, fragments of arithmetic, reverse mathematics, theory of algorithmic randomness and computation theory of the reals. He also has a unified view of mathematical logic and of its role in mathematics, for example, he has worked on computations applied to complex dynamics and is interested in applying recursion theory to number theory and geometry.

1.1. *Admissible recursion theory*

In the 1960's Kreisel and Sacks generalized computability to admissible ordinals. Between 1973, when starting his mathematical career, and 1984, when finishing his influential Springer monograph "Techniques of Admissible Recursion Theory",[4] this area was CT's key interest. His paper "Generic sets and minimal α-degrees"[3] showed that any construction of a minimal α-degree was reducible to Sacks-Spector forcing. The paper shed light on what is possible and necessary for any such construction, and is highlighted in Sacks's classical book "Higher Recursion Theory".[38] 20 years later, he and Slaman established the undecidability of the theory of α-degrees.[17] This has been an open problem for a long time mainly due to the inadequacy of classical tools. They overcame this difficulty by an ingenuous use of Slaman-Woodin coding, α-finite injury method and forcing.

Joint with Sy D. Friedman, CT contributed an article "Ordinal recursion theory" in the Handbook of Computability Theory.[9] According to the Math. Review, "This paper is a state-of-the-art exposition of modern α-recursion theory and β-recursion theory".

1.2. *Recursion theory over fragments of Peano arithmetic*

In the 1980's Slaman and Simpson studied computability on nonstandard models of weak fragments of arithmetic. This generalization of computability immediately attracted CT's attention, because in the fragments of arithmetic, Σ_n-induction played a similar role as Σ_n-admissibility on ordinals. In paper,[14] he and Mourad proved Friedberg-Muchnik Theorem without assuming Σ_1-induction, which was particularly challenging since no priority construction was possible in this situation. In a series of papers,[16,22–24] CT and other collaborators classified the infinite injury priority constructions in terms of the strength of induction. This elucidates the nature of recursion-theoretic constructions in relation to proof-theoretic strength of statements about Turing degrees. Questions of this nature were raised earlier by Shoenfield. The logic community used to view Shoenfield's question as more of a philosophical nature, perhaps impossible to answer. It is generally believed that the priority constructions should have something to do with the syntactical form of the set of requirements that the construction needed to satisfy. However, the way of writing requirements is far from unique. Using the strength of induction to classify the priority constructions, it not only confirm the intuitive connection between the construction and the aforementioned syntactical form, but also independent of the choice of requirements. Finding the exact strength of induction required has a strong flavor of reverse mathematics, which will be discussed in Section 2.

1.3. *Classical recursion theory*

Of course, every recursion theorist is interested in the Turing degree structures, local or global. CT is no exception. In,[10] he and Jockusch showed that every nonzero recursively enumerable degree bounds a 1-generic degree, which was the first work on the structure of 1-generic degrees below $0'$. Subsequently he and Downey[7,8] made two other contributions on 1-generic degrees. This was a precursor to Kumabe's comprehensive work in 1-generic degrees.[30,31]

Occasionally, CT also worked on some local degree problems purely out of technical interest. For example, in "The existence of high nonbounding degrees in the difference hierarchy", CT, Angsheng Li and Yue Yang proved that there is a high d-r.e. degree d which does not bound any minimal pair of d-r.e. degrees. As a corollary they also proved that there is a high d-r.e. degree which bounds no Slaman triple (a, b, c), i.e. degrees a, b, c, such that $0 < a$, and $c \not\leq b$, and for every w from $(0, a)$, $c \leq b \lor w$. This result contrasts with a known result of Shore and Slaman[40] that any high r.e. degree bounds a Slaman triple. (A Slaman triple in r.e. degrees remains a Slaman triple in d-r.e. degrees.)

Recently, CT also worked on global Turing degrees structures, a research area at the intersection of recursion theory and set theory. Joint with Liang Yu, they studied maximal chains and antichains in Turing degrees. A maximal chain of Turing degrees is a set of reals which are pairwise Turing comparable, but not equivalent. The existence of a maximal chain follows from Zorn's lemma. In "Maximal chains in the Turing degrees",[25] CT and L. Yu studied whether the use of the axiom of choice can be removed, and whether there exists a definable, say Π_1^1 maximal chain. They showed that the system ZF +DC +"There exists no maximal chain of Turing degrees" is equiconsistent with the system ZFC+"There is an inaccessible cardinal". Another example is the study of Martin's conjecture, which is one of the most important open questions in the area. In "The strength of the projective Martin conjecture",[21] CT, Wei Wang and L. Yu studied Martin's Conjecture on Π_1^1-functions uniformly $\leq_T$-order preserving on a cone implies Π_1^1 Turing Determinacy over ZF +DC. In addition, they also proved that for $n \geq 0$, this conjecture for uniformly degree invariant Π_{2n+1}^1-functions is equivalent over ZFC to Σ_{2n+2}^1-Axiom of Determinacy. As a corollary, the consistency of the conjecture for uniformly degree invariant Π_1^1 functions implies the consistency of the existence of a Woodin cardinal.

In 2015, CT and L. Yu published a monograph on higher recursion theory and their applications.[26] To quote Ted Slaman in Math Review: "This book serves two purposes, and does so very well. In Part I, it provides an exposition of the now-classical theory of definability in first-order arithmetic, in the form of the arithmetic hierarchy, and in second-order arithmetic, in the form of effective descriptive set theory. This part would be a good source on which to base a graduate course on this material. In Parts II-IV, by giving a coherent and systematic treatment spanning many modern examples, it illustrates how these classical ideas have evolved into powerful mathematical tools, which is valuable both to newcomers and to experts."

1.4. *Applications of Recursion Theory*

There are still two areas in applied Recursion Theory that CT has made important contributions. One is algorithmic randomness which is arguably an area within Recursion Theory after about 20 years of interaction. In,[15] he with Nies and Yu demonstrated that methods of effective descriptive set theory are fruitful tools for problems in higher randomness. More precisely, they established the relationships Π_1^1-randomness $\subset \Pi_1^1$-Martin-Löf randomness $\subset \Delta_1^1$-randomness $= \Delta_1^1$-Martin-Löf randomness. They also characterized the set of reals that are low for Δ_1^1- randomness as precisely those that are Δ_1^1-traceable.

The other area is computation theory over the real and Complex numbers. After Blum, Shub and Smale introduced their computation model over arbitrary rings in 1989,[1] CT was one of the first to study the degree of insolvability of filled Julia sets. In,[5] he studied Julia sets of polynomials under recursion-theoretic aspects. Using the notion of decidability refers to the real number model introduced by Blum, Shub and Smale, he established the following result: Let f be a complex polynomial. Consider its Julia set $J(f)$ and the set $K(f)$ of points not tending to infinity under iteration of f. For a certain class of polynomials, the interior of $K(f)$ is undecidable as well as strictly reducible to $J(f)$.

2. CT's Work in Ramsey-Type Theorems and Reverse Mathematics

2.1. *Ramsey's Theorem*

In layman's terms, Ramsey's Theorem for pairs can be stated as follows: In a party of infinitely many participants, we can always find an infinite group of people such that all of them knew each other or all of them are mutual strangers. The precise statement of the general version of Ramsey's Theorem says:

Theorem 2.1 (Ramsey[37]). *Any $f : [\mathbb{N}]^n \to \{0, 1, \ldots, k - 1\}$ has an infinite homogeneous set $H \subseteq \mathbb{N}$, namely, f is constant on $[H]^n$; here $[X]^n$ stands for the set of all n-element subsets of X.*

If we think of f as a k-coloring of the n-element subsets of natural numbers, then there is an infinite set H, whose n-element subsets have the same color. It is customary to think of this kind of "for all f there exists H..." statements as "our opponent posed a problem (e.g. colouring) and we must provide a solution (e.g., the infinite homogenous set)".

The version above is denoted by RT^n_k. Our main focus is on RT^2_2 — Ramsey's Theorem for Pairs. We now sketch a proof of RT^2_2. Let f be a coloring of pairs, say by red and blue. We first find an infinite subset C of natural numbers on which f is "stable", i.e., for all $x \in C$, the limit $\lim_{y \in C} f(x, y)$ exists. We call such a set C *cohesive for f*. Next we consider the following two sets: $D^R = \{x \in C : x \text{ is "eventually red"}\}$ and $D^B = \{x \in C : x \text{ is "eventually blue"}\}$. One of them must be infinite, say it is D^R. Now it is fairly easy to select the elements of a red homogeneous set from D^R: Let a_0 be the least element in D^R. Suppose that we have selected $a_0 < a_1 < \cdots < a_k$, let a_{k+1} be the first element larger than a_k such that (a_i, a_{k+1}) is coloured red for all $i \le k$. The existence of a_{k+1} is guaranteed by the stability.

We extract two combinatorial principles out of the proof: Let R be an infinite subset of natural numbers and $R_s = \{t | \langle s, t \rangle \text{ is in } R\}$ where $\langle s, t \rangle$ stands for the Gödel coding of pairs. A set G is said to be R-cohesive if for all s, either $G \cap R_s$ is finite or $G \cap (\mathbb{N} \setminus R_s)$ is finite. The cohesive principle COH states that for every R, there is an infinite G that is R-cohesive. The other principle is called the stable Ramsey's Theorem for pairs, denoted by SRT_2^2, which states that every stable coloring of pairs has a solution. The principles COH and SRT_2^2 were studies by Cholak, Jockusch and Slaman,[2] where they showed

Theorem 2.2 (Cholak, Jockusch and Slaman).
$$\mathsf{RT}_2^2 = \mathsf{SRT}_2^2 + \mathsf{COH}.$$

There are many other principles which are corollaries of Ramsey's Theorem for pairs. For instance, the principle ADS of ascending or descending sequence states that every infinite linearly ordered set contains an infinite subsequence that is either increasing or decreasing. The Chain and Antichain Principle CAC states that every infinite partially ordered set has an infinite chain or antichain. However, the decomposition in Theorem 2.2 turns out to be extremely useful when people wanted to break RT_2^2 and to apply the divide and conquer method.

2.2. *Reverse Mathematics*

It is natural to study the relative strength of combinatorial principles, in particular, the principles related to Ramsey's theorem. It turns out that the most interesting ones are those weaker than the Ramsey's theorem for pairs. For example, it is natural to ask whether COH or SRT_2^2 as strong as RT_2^2? and whether ADS implies RT_2^2? Or to put it more generally, what are the logical consequences/strength of a combinatorial principle, for example, Ramsey's Theorem?

To answer these questions, one needs to bring in logical tools. For example, the answer may depend on analyzing the complexity of the homogeneous set H. Also, one needs logic to determine if one principle P implies the other principle Q. It is usually more challenging to show that P does not imply Q. As we know from logic that one way to demonstrate that $P \nRightarrow Q$ is to "make" P true and Q false. But these combinatorial principles are all true theorems from mathematics, how can one make some of them false?

Thus we have to work in some weaker axiomatic system Γ, and demonstrate that "Γ proves P but does not prove Q". Usually, we will have a hierarchy of axiomatic systems $\Gamma_0 < \Gamma_1 < \dots$ as our benchmarks and their relative strength has been established that Γ_i is strictly weaker than Γ_j for $i < j$ (as indicated by the less than symbol between the systems). Therefore, to show that the P does not prove Q, it suffices to show that Γ_i proves P and on the other hand Q proves Γ_j for some $j > i$. Notice that the last step requires that we prove axiom Γ_j from a theorem Q, which reverses the usual mathematical practice of proving theorems from axioms, that is where the name "reverse mathematics" comes. Another advantage of using logical systems is that we could establish a large class of results across different areas of mathematics, rather than just comparing the strength of two special instances P and Q. The standard reference book for reverse mathematics is Simpson.[41]

We now introduce two most commonly used axiomatical systems for that purpose, namely, the subsystems of first- and second-order arithmetic. Recall that the language of first order Peano Arithmetic contains a constant symbol 0, three function symbols $S, +, \times$, and a binary predicate $<$. Formulas over the language of arithmetic naturally form a hierarchy by the number of alternating blocks of quantifiers, which gives us the usual arithmetic hierarchy. Formulas with n alternating blocks of quantifiers with leading one existential (or universal) are called Σ_n^0 and Π_n^0 respectively. The superscript 0 is to indicate the formulas are first-order. For example, $(\forall x, y, z > 2)[x^3 + y^3 \neq z^3]$ is Π_1^0 (here x^3 is just a shorthand for the product of $x \cdot x \cdot x$). Furthermore, the Δ_n^0 formulas are those having two equivalent forms, one Σ_n^0 and Π_n^0. Let $I\Sigma_n^0$ denote the induction schema for Σ_n^0-formulas; and $B\Sigma_n^0$ denote the Bounding Principle for Σ_n^0-formulas. By a theorem of Kirby and Paris[36]

$$\dots \Rightarrow I\Sigma_{n+1}^0 \Rightarrow B\Sigma_{n+1}^0 \Rightarrow I\Sigma_n^0 \Rightarrow \dots$$

we have one benchmark in first-order arithmetic.

The other benchmark is by subsystems of second order arithmetic which is used in reverse mathematics. In second order arithmetic, the variables and quantifiers can range over sets or relations. For example, "every nonempty subset has a least element". Complexity can be defined similarly, for example, the above statement is Π_1^1. Here we only list three of those subsystems which are needed in the sequel: RCA$_0$ which contains Σ_1^0-induction and Δ_1^0-comprehension: For any Δ_1^0-formula φ, $\exists X \forall n (n \in X \leftrightarrow \varphi(n))$; WKL$_0$ which is RCA$_0$ plus weak König Lemma saying that every infinite binary tree has an infinite path; and ACA_0 which is RCA_0 plus arithmetical comprehension. Their relative strength are known:

$$RCA_0 < WKL_0 < ACA_0.$$

We also need the notion of models. A model $\mathcal{M}$ of second-order arithmetic consists $(M, 0, S, +, \times, <, X)$, where $(M, 0, S, +, \times, <)$ is its first-order part and the set variables are interpreted as members of X. For example, if $\mathcal{M}$ is a model of RCA_0, then its second-order part X is closed under Turing reducibility and Turing join. In short, we have two measures of strength: first-order measure which is by the amount of induction required and second order measure which is by the richness of set existence.

With the concept of hierarchies available, we can further rephrase the motivating questions: Suppose the coloring function f is recursive, what is the minimal syntactical complexity of a solution? Which system in Reverse Mathematics does Ramsey's Theorem correspond? E.g., does RT_2^2 imply ACA_0? What are the first-order consequences of Ramsey's Theorem? E.g., does RT_2^2 imply $I\Sigma_2^0$? Does SRT_2^2 imply RT_2^2? More precisely, if a model of RCA_0 whose second order part X contains solutions for all stable colorings in X, must X contain solutions for all colorings in X?

2.3. Earlier results

We now give a list of historical results. Some of the early studies are motivated by effective mathematics, we modified their statements to suit our purposes.

Theorem 2.3 (Jockusch[32]). *Over* RCA_0,

$$ACA_0 \Leftrightarrow RT_2^3 \Leftrightarrow RT_k^n$$

where $n, k \geq 3$ *and*

$$ACA_0 \Rightarrow RT_2^2 \quad and \quad WKL_0 \not\Rightarrow RT_2^2.$$

Theorem 2.4 (Hirst[29]). *Over* RCA_0,

$$(S)RT_2^2 \Rightarrow B\Sigma_2.$$

This tells us a lower bound of its first-order strength.

Theorem 2.5 (Seetapun and Slaman[39]). *Over* RCA_0,

$$RT_2^2 \not\Rightarrow ACA_0.$$

Seetapun's proof made clever use of trees, which leads to Seetapun Conjecture that $RT_2^2 \Rightarrow WKL_0$.

To determine an upper bound of first-order strength, conservation results are often used. One of the pioneer conservation results is done by Harrington, who showed that WKL_0 is Π_1^1-conservative over RCA_0, i.e., any Π_1^1-statement that is provable in WKL_0 is already provable in RCA_0 (see Corollary IX.2.6 in[41]).

Theorem 2.6 (Cholak, Jochusch and Slaman[2]). RT_2^2 is Π_1^1-conservative over $RCA_0 + I\Sigma_2$.

Corollary 2.1 (Cholak, Jochusch and Slaman[2]). Over RCA_0,

$$RT_2^2 \not\vdash I\Sigma_3.$$

2.4. *Making progress on* RT_2^2

After Cholak, Jockusch and Slaman's paper, the exact strength of RT_2^2 was studied extensively by practically every expert in the field and many failed attempts were made to solve it. However, the extensive study changed the whole field of reverse mathematics, for example, the usual subsystems are no longer the only benchmarks to use, in fact, around the RT_2^2, linear measurement is no longer sufficient, it is more like a "zoo" now. Hirschfeldt and Shore[28] made further progress on the exact strength of many important combinatorial principles weaker than RT_2^2, for instance, they showed that ADS is strictly weaker than RT_2^2. However, three major questions remained open at that time: (1) Seetapun's Conjecture; (2) Over RCA_0, does SRT_2^2 imply RT_2^2? (3) Does SRT_2^2 imply $I\Sigma_2$? If not, how about RT_2^2?

I would add here that Hirschfeldt and Shore[28] also asked many questions directly related to fragments of arithmetic, which was actively investigated by NUS logic group as mentioned above. In fact, CT had investigated reverse mathematics and combinatorial principles related to Ramsey's Theorem since 2003. In,[12] CT, Steffen Lempp and Yue Yang elaborated on the role of the bounding principle $B\Sigma_2^0$ in second-order reverse mathematics. They introduced the notion of a bi-tame cut in a model of Σ_1^0-induction, and showed that such a model is a model of $B\Sigma_2^0$ if and only if it has no bi-tame cuts. They used this result to answer a question of Hirschfeldt and Shore[28] by showing that the principle PART introduced in that paper is equivalent to $B\Sigma_2^0$ over RCA_0. They also filled a gap in a proof of Cholak, Jockusch and Slaman[2] and showed that the principle D_2^2 introduced in that paper is equivalent to SRT_2^2 over RCA_0. In,[18] CT, Ted Slaman and Yue Yang showed that the principles COH, CAC and ADS are all Π_1^1-conservative over $RCA_0 + B\Sigma_2^0$. The proof of the conservation of COH is technically innovative, as it requires two-step forcing, an internal one which is within a given model, followed by an external forcing, which made use of countability of the model.

Now, back to the three major problems above. The first problem was solved by Jiayi Liu (aka Lu Liu),[33] he showed that

Theorem 2.7 (Jiayi Liu[33]). *Over* RCA_0,

$$RT_2^2 \not\Rightarrow WKL_0.$$

However, the solution for (2) and (3) remained elusive. The most natural approach is to show that stable colorings always have a low solution (here the word "low" is a technical term in recursion theory). Or equivalently, every Δ_2^0-set contains or is disjoint from an infinite low set. However, Downey, Hirschfeldt, Lempp and Solomon[27] showed that there is a Δ_2^0 set D such that neither D nor $\overline{D}$ contains infinite low subset, thus blocked the seemingly only promising approach.

It is this place that CT made the crucial contribution. Using his experience on nonstandard models of arithmetic, he in[6] suggested that we should look at nonstandard models of fragments of arithmetic, because the theorem by Downey, Hirschfeldt, Lempp and Solomon was done on the standard model of arithmetic, whose proof involves infinite injury method thus requires $I\Sigma_2$. Yet we know that in nonstandard models things behave differently, for example, there is a model of $B\Sigma_2$ but not $I\Sigma_2$ in which every incomplete Δ_2^0 set is low. In the end, this approach turned out to be fruitful and question (2) and (3) were answered:

Theorem 2.8 (Chong, Slaman and Yang[19]). *Over* RCA_0,

$$SRT_2^2 \not\Rightarrow RT_2^2$$

$$SRT_2^2 \not\Rightarrow I\Sigma_2.$$

Theorem 2.9 (Chong, Slaman and Yang[20]).

$$RT_2^2 \not\Rightarrow I\Sigma_2.$$

Recently, Monin and Patey[34] announced that SRT_2^2 does not imply RT_2^2 in ω-models.

I would like to end the paper by mentioning some ongoing projects related to Ramsey's Theorem on trees. Inspired by RT_k^n, one can take a global view and consider a general program, termed "structural Ramsey theory" (Todorcevic,[42] Nguyen[35]), that investigates the phenomenon of Ramsey-type homogeneity in a mathematical structure. Since 2018, CT and his collaborators studied the reverse mathematics of the combinatorial principle TT_k^n derived from the tree theorem in structural Ramsey theory. This theorem states that for each n and k, any k-coloring of comparable tuples produces an isomorphic homogeneous subtree, i.e. a tree all of whose compatible n-tuples have the same color. In,[11] CT, Wei Li, Wei Wang and Yue Yang showed that over the system RCA_0, the inductive strength of TT^1, which is $\forall k TT_k^1$, is weaker than Σ_2^0-induction, which followed from the main theorem in the paper saying that over the same system, TT^1 is Π_1^1-conservative over Σ_2^0-bounding $+ P\Sigma_1^0$, where $P\Sigma_1^0$ is the principle equivalent to the totality of the Ackermann function. In,[13] CT, Wei Li, Lu Liu and Yue Yang showed that over RCA_0, TT_k^2 does not imply WKL_0, thus solved the open problem on the relative strength between the two major subsystems in second order arithmetic. These projects about Ramsay's Theorem on trees are still being carried out, we expect that more results will come in the near future.

References

1. L. Blum, M. Shub, and S. Smale. On a theory of computation over the real numbers: NP completeness, recursive functions and universal machines. *Bull. Amer. Math. Soc.* (N.S.) 21(1): 1–46, 1989.

2. Peter A. Cholak, Carl G. Jockusch, and Theodore A. Slaman. On the strength of Ramsey's theorem for pairs. *J. Symbolic Logic*, 66(1): 1–55, 2001.

3. C. T. Chong. Generic sets and minimal α-degrees. *Trans. Amer. Math. Soc.*, 254(1): 157–169, 1979.

4. C. T. Chong. *Techniques of Admissible Recursion Theory.* Lecture Notes in Mathematics, Vol. 1106, Springer Verlag, 1984.

5. C. T. Chong. Positive reducibility of the interior of filled Julia sets. *J. Complexity*, 10(4): 437–444, 1994.

6. C. T. Chong. Nonstandard methods in Ramsey's Theorem for pairs. in *Computational Prospects of Infinity, Part II: Presented Talks.* World Scientific, 47–58, 2006.

7. C. T. Chong, and R. G. Downey. Degrees bounding minimal degrees. *Math. Proc. Cambridge Philos. Soc.*, 105(2): 211–222, 1989.

8. C. T. Chong, and R. G. Downey. Minimal degrees recursive in 1-generic degrees. *Ann. Pure Appl. Logic*, 48(3): 215–225, 1990.

9. C. T. Chong, and Sy. D. Friedman. Ordinal recursion theory. in *Handbook of Computability Theory*, 277–299, Stud. Logic Found. Math., 140, North-Holland, Amsterdam, 1999.

10. C. T. Chong, and Carl G. Jockusch. Minimal degrees and 1-generic sets below $0'$. *Computation and proof theory (Aachen, 1983)*, 63–77, Lecture Notes in Math., 1104, Springer, Berlin, 1984.

11. C. T. Chong, W. Li, W. Wang and Y. Yang. On the strength of Ramsey's theorem for trees. *Adv. Math.*, 369 (2020), 107180, 39 pp.

12. C. T. Chong, Steffen Lempp, and Yue Yang, On the role of the collection principle for Σ_2^0-formulas in second-order reverse mathematics. *Proc. Amer. Math. Soc.*, 138(3): 1093–1100, 2010.

13. C. T. Chong, W. Li, L. Liu, and Y. Yang, The Strength of Ramsey's Theorem For Pairs over trees: I. Weak König's Lemma. to appear in *Trans. Amer. Math. Soc.*

14. C. T. Chong and K. J. Mourad. Σ_n definable sets without Σ_n induction. *Trans. Amer. Math. Soc.*, 334(1): 349–363, 1992.

15. C. T. Chong, A. Nies, and L. Yu. Lowness of higher randomness notions. *Israel J. Math.*, 166, 39–60, 2008.

16. C. T. Chong, Lei Qian, Theodore A. Slaman, and Yue Yang, Σ_2 induction and infinite injury priority arguments, Part III: Prompt sets, minimal pair and Shoenfield's conjecture. *Israel J. Math.*, 121(1): 1–28, 2001.

17. C. T. Chong, and Theodore A. Slaman. The theory of α-degrees is undecidable. *Israel J. Math.*, 178, 229–252, 2010.

18. C. T. Chong, Theodore A. Slaman, and Yue Yang. Π_1^1-conservation of combinatorial principles weaker than Ramsey's theorem for pairs. *Adv. Math.*, 230(3): 1060–1077, 2012.

19. C. T. Chong, Theodore A. Slaman, and Yue Yang. The metamathematics of stable Ramsey's theorem for pairs. *J. Amer. Math. Soc.*, 27(3): 863–892, 2014.

20. C. T. Chong, Theodore A. Slaman, and Yue Yang. The inductive strength of Ramsey's theorem for pairs, *Adv. Math.*, 308, 121–141, 2017.

21. C. T. Chong, Wei Wang, and Liang Yu. The strength of the projective Martin conjecture. *Fund. Math.*, 207(1): 21–27, 2010.

22. C. T. Chong, and Yue Yang, Σ_2 induction and infinite injury priority arguments, part II: Tame Σ_2 coding and the jump operator. *Ann. Pure Appl. Logic*, 87(2): 103–116, 1997.

23. C. T. Chong, and Yue Yang. Σ_2 induction and infinite injury priority arguments, part I: Maximal sets and the jump operator. *J. Symbolic Logic*, 63(3): 797–814, 1998.

14

24. C. T. Chong, and Yue Yang. The jump of a Σ_n-cut. *J. Lond. Math. Soc.*, 75(3): 690–704, 2007.
25. C. T. Chong, and Liang Yu. Maximal chains in the Turing degrees. *J. Symbolic Logic*, 72(4): 1219–1227, 2007.
26. C. T. Chong, and Liang Yu. *Recursion theory. Computational aspects of definability. With an interview with Gerald E. Sacks.* De Gruyter Series in Logic and its Applications, 8. De Gruyter, Berlin, 2015.
27. Rod Downey, Denis R. Hirschfeldt, Steffen Lempp, and Reed Solomon. A Δ_2^0 set with no infinite low subset in either it or its complement. *J. Symbolic Logic*, 66(3): 1371–1381, 2001.
28. Denis Hirschfeldt, and Richard Shore. Combinatorial principles weaker than Ramsey's theorem for pairs. *J. Symbolic Logic*, 72(1): 171–206, 2007.
29. J. L. Hirst. *Combinatorics in Subsystems of Second Order Arithmetic.* PhD thesis, The Pennsylvania State University, 1987.
30. Masahiro Kumabe. Every n-generic degree is a minimal cover of an n-generic degree. *J. Symbolic Logic*, 58(1): 219–231, 1993.
31. Masahiro Kumabe. A 1-generic degree with a strong minimal cover. *J. Symbolic Logic*, 65(3): 1395–1442, 2000.
32. Carl G. Jockusch, Jr. Ramsey's theorem and recursion theory. *J. Symbolic Logic*, 37: 268–280, 1972.
33. Jiayi Liu. RT_2^2 does not prove WKL_0. *J. Symbolic Logic*, 77(2): 609–620, 2012.
34. B. Monin and L. Patey. SRT_2^2 does not imply COH in ω-models. arXiv preprint, arXiv:1905.08427, 2019.
35. Nguyen Van Thé, Lionel. A survey on structural Ramsey theory and topological dynamics with the Kechris-Pestov-Todorcevic correspondence in mind. in: *Selected Topics In Combinatorial Analysis.* Zbornik Radova (Beograd) 17(25): 189–207, 2015.
36. J. B. Paris and L. A. S. Kirby. Σ_n-collection schemas in arithmetic. In *Logic Colloquium '77 (Proc. Conf., Wrocław, 1977)*, volume 96 of *Stud. Logic Foundations Math.*, pages 199–209. North-Holland, Amsterdam, 1978.
37. F. P. Ramsey. On a Problem of Formal Logic. *Proc. London Math. Soc.* (2) 30(4): 264–286, 1929.
38. Gerald E. Sacks. *Higher Recursion Theory.* Perspectives in Mathematical Logic, vol. 2, Springer–Verlag, Berlin, Heidelberg, 1990.
39. David Seetapun and Theodore A. Slaman. On the strength of Ramsey's theorem. *Notre Dame J. Formal Logic*, 36(4): 570–582, 1995.
40. Richard A. Shore, and Theodore A. Slaman. Working below a high recursively enumerable degree. *J. Symbolic Logic*, 58(3): 824–859, 1993.
41. Stephen G. Simpson. *Subsystems of second order arithmetic.* Perspectives in Logic. Cambridge University Press, Cambridge, second edition, 2009.

42. Stevo Todorcevic. *Introduction to Ramsey Spaces*. Annals of Mathematics Studies, 174. *Princeton University Press*, Princeton, NJ, 2010. viii+287 pp.

On One-Variable Fragments of Modal μ-Calculus

Leonardo Pacheco

Mathematical Institute, Tohoku University,
Sendai, 9808578, Japan
E-mail: leonardovpacheco@gmail.com

Wenjuan Li

Division of Mathematical Sciences, Nanyang Technological University,
Singapore 637371
E-mail: wenjuanli1701@gmail.com

Kazuyuki Tanaka

Mathematical Institute, Tohoku University,
Sendai, 9808578, Japan
E-mail: tanaka.math@tohoku.ac.jp

In this paper, we study one-variable fragments of modal μ-calculus and their relations to parity games. We first introduce the weak modal μ-calculus as an extension of the one-variable modal μ-calculus. We apply weak parity games to show the strictness of the one-variable hierarchy as well as its extension. We also consider games with infinitely many priorities and show that their winning positions can be expressed by both Σ_2^μ and Π_2^μ formulas with two variables, but requires a transfinite extension of the L_μ-formulas to be expressed with only one variable. At last, we define the μ-arithmetic and show that a set of natural numbers is definable by both a Σ_2^μ and a Π_2^μ formula of μ-arithmetic if and only if it is definable by a formula of the one-variable transfinite μ-arithmetic.

Keywords: μ-calculus; Parity Games; Difference Hierarchy; μ-arithmetic.

1. Introduction

Modal μ-calculus is an extension of modal logic by adding greatest and least fixpoint operators. By modal logic, we mean the propositional logic with modalities $\Box$ (universal modality, which is interpreted as necessity) and $\Diamond$ (existential modality, which is interpreted as possibility). Modal μ-calculus has been extensively investigated from automata-theoretic and game-theoretic viewpoints, especially in the relationships among modal μ-calculus, (alternating) tree automata and parity games.[15,17]

A fundamental issue on modal μ-calculus is the strictness of *alternation hierarchy*. The alternation hierarchy classifies formulas by their *alternation depth*, namely, the number of alternating appearances of greatest and least fixpoint operators. Note that the alternation depth, in a game-theoretic view, is related with the number of priorities in parity games, and from an automata-theoretic perspective, it concerns with the Rabin index (or parity index) of tree automata.

The strictness of alternation hierarchy of modal μ-calculus was established by Bradfield,[10-12] and at the same time by Lenzi.[21] In sequel, Arnold[3] and Bradfield[9] showed that the alternation hierarchy of modal μ-calculus is strict over infinite binary trees. Alberucci and Facchini[1] proved that the alternation hierarchy is strict over reflexive transition systems. D'Agostino and Lenzi[13] further showed the hierarchy of the modal μ-calculus over reflexive and symmetric graphs is infinite.

At the same time, the number of variables contained in a formula also serves as an important measure of complexity for formulas of modal μ-calculus. Berwanger[4] proved that the two-variable fragment of modal μ-calculus is enough to express properties in arbitrary level of alternation hierarchy of modal μ-calculus. Then it is natural to ask if $L_\mu[2] = L_\mu$ and the distinct bound variables are redundant. Berwanger, Grädel and Lenzi[5,6] gave a negative answer to this question by showing the strictness of variable hierarchy of modal μ-calculus.

Similar to the addition of inductive definition to modal logic, Lubarsky[24] first introduced fixpoint constructors to first-order arithmetic and obtained the so-called μ-arithmetic, which makes a great increase in the expressive power. Bradfield[10,12] used it to show the strictness of the alternation hierarchy of modal μ-calculus, by showing that the set of natural numbers defined by the alternation hierarchy of μ-arithmetic can also be characterized by the alternation hierarchy of modal μ-calculus over *recursively presentable transition systems*. When extending to transfinite hierarchies, Bradfield[7] applied parity games constructed from the difference hierarchy of Σ_2^0 sets.

In this paper, we study the one-variable fragment of modal μ-calculus and an extension called weak modal μ-calculus. The latter is essentially the same as $\Sigma_2^\mu \cap \Pi_2^\mu$ in the alternation hierarchy due to Emerson-Lei.[16] Although this result was already proved by Mostowski,[27] we will give a proof by using weak parity games. Finally, we introduce the weak parity games with infinitely many priorities. We show that their winning regions can be expressed by a Σ_2^μ formula and a Π_2^μ, but in term of one-variable fragment, it requires a transfinite extension of alternation hierarchy of one-variable modal μ-calculus. To this end, we introduce the transfinite weak μ-arithmetic and show its connection to the transfinite one-variable modal μ-calculus.

The remainder of this paper is organized as follows. In section 2, we recall the definitions of basic concepts about modal μ-calculus and parity games, and some classical results on the relationships among them. In section 3, we show that a weak parity game can be a witness for the strictness of one-variable hierarchy. In section 4, we introduce a transfinite extension of one-variable modal μ-calculus. In section 5, we introduce the transfinite weak μ-arithmetic, and explain the connection between the transfinite one-variable μ-arithmetic and weak parity games.

2. Preliminaries

2.1. *Modal μ-calculus*

2.1.1. *Syntax*

To start with, we define the formulas of modal μ-calculus as follows.

Definition 2.1 (L_μ-formula). *Given a language consisting of*

- $\mathfrak{P}$: *a set of atomic propositions, P, Q, ...*
- $\mathfrak{X}$: *a countable set of second-order variables, X, Y, ...*

the set L_μ of formulas of modal μ-calculus is inductively defined as follows:

$$\varphi := \top \mid \bot \mid P \mid \neg P \mid X \mid \varphi \wedge \varphi \mid \varphi \vee \varphi \mid \Diamond\varphi \mid \Box\varphi \mid \mu X.\varphi \mid \nu X.\varphi$$

where $P \in \mathfrak{P}$ and $X \in \mathfrak{X}$. Here, μX and νX are called the least and greatest fixpoint operators.

The *negation* of a formula is introduced by the following rules:

- $\neg(\neg X) = X$,
- $\neg(\psi \vee \varphi) = \neg\psi \wedge \neg\varphi$,
- $\neg(\psi \wedge \varphi) = \neg\psi \vee \neg\varphi$,
- $\neg\Box\varphi = \Diamond\neg\varphi$,
- $\neg\Diamond\varphi = \Box\neg\varphi$,
- $\neg\mu X.\varphi(X) = \nu X.\neg\varphi(\neg X)$,
- $\neg\nu X.\varphi(X) = \mu X.\neg\varphi(\neg X)$.

Notice that only positive occurrences of a variable X are allowed in a formula, namely an even number of negations nesting X. Such a condition ensures the existence of the least and greatest fixpoints.

We use the symbol η to denote either μ or ν. An L_μ-formula of the form $\mu X.\varphi$ is called a *μ-formula*, and the form $\nu X.\varphi$ is called a *ν-formula*. Given a formula, (the occurrence of) a variable X is said to be bound if it occurs in some subformula of the form $\eta X.\varphi$. If (the occurrence of) X is not bound, then it is said to be free. A formula with no free (occurrence of) variables is called a *sentence*.

2.1.2. *Denotational semantics*

A sentence $\varphi \in L_\mu$ can be evaluated via *pointed transition systems* $\mathbb{S} = (S, \llbracket\ \rrbracket^\mathbb{S})$ (also called *Kripke models*) at a particular vertex $s \in S$, in the sense that

$$(\mathbb{S}, s) \models \varphi \overset{def}{\Longleftrightarrow} s \in \llbracket\varphi\rrbracket^\mathbb{S}.$$

More formally, a transition system is a tuple $\mathbb{S} = (S, R, \mathcal{I})$, where

- S is a set of states (or vertices),
- $R \subseteq S \times S$ is a transition relation, and
- $\mathcal{I} : \mathfrak{P} \to \mathcal{P}(S)$ is a valuation.

A valuation can be regarded as a labeling on the states, that is, each state $s \in S$ is labelled by $\{P \in \mathfrak{P} : s \in \mathcal{I}(P)\} \subseteq \mathfrak{P}$.

Given a transition system $\mathbb{S}$ and a valuation function $\mathcal{V} : \mathfrak{X} \to \mathcal{P}(S)$, we define $[\![\cdot]\!]_{\mathcal{V}}^{\mathbb{S}} \subseteq S$ as follows:

$$[\![\top]\!]_{\mathcal{V}}^{\mathbb{S}} := S \text{ and } [\![\bot]\!]_{\mathcal{V}}^{\mathbb{S}} := \emptyset,$$

$$[\![X]\!]_{\mathcal{V}}^{\mathcal{M}} := \mathcal{V}(X) \quad \text{for all } X \in \mathfrak{X},$$

$$[\![P]\!]_{\mathcal{V}}^{\mathcal{M}} := \mathcal{I}(P) \text{ and } [\![\neg P]\!] := S \setminus \mathcal{I}(P) \quad \text{for all } P \in \mathfrak{P},$$

$$[\![\varphi \vee \psi]\!]_{\mathcal{V}}^{\mathbb{S}} := [\![\varphi]\!]_{\mathcal{V}}^{\mathbb{S}} \cup [\![\psi]\!]_{\mathcal{V}}^{\mathbb{S}},$$

$$[\![\varphi \wedge \psi]\!]_{\mathcal{V}}^{\mathbb{S}} := [\![\varphi]\!]_{\mathcal{V}}^{\mathbb{S}} \cap [\![\psi]\!]_{\mathcal{V}}^{\mathbb{S}},$$

$$[\![\Diamond\varphi]\!]_{\mathcal{V}}^{\mathbb{S}} := \left\{v \in S : \exists w, (v, w) \in R \wedge w \in [\![\varphi]\!]_{\mathcal{V}}^{\mathbb{S}}\right\},$$

$$[\![\Box\varphi]\!]_{\mathcal{V}}^{\mathbb{S}} := \left\{v \in S : \forall w, (v, w) \in R \to w \in [\![\varphi]\!]_{\mathcal{V}}^{\mathbb{S}}\right\},$$

$$[\![\mu X.\varphi]\!]_{\mathcal{V}}^{\mathbb{S}} := \bigcap\left\{S' \subseteq S : [\![\varphi]\!]_{\mathcal{V}[X:=S']}^{\mathbb{S}} \subseteq S'\right\},$$

$$[\![\nu X.\varphi]\!]_{\mathcal{V}}^{\mathbb{S}} := \bigcup\left\{S' \subseteq S : [\![\varphi]\!]_{\mathcal{V}[X:=S']}^{\mathbb{S}} \supseteq S'\right\}.$$

Finally, if φ is a sentence, we may write $[\![\varphi]\!]^{\mathbb{S}}$ for $[\![\varphi]\!]_{\mathcal{V}}^{\mathbb{S}}$, since its value is independent from $\mathcal{V}$.

2.1.3. Game semantics

The semantics of modal μ-calculus can be also defined in terms of evaluation games. The two players, $\exists$ and $\forall$, move on a graph generated by the given L_μ-formula and a transition system. We say that player $\exists$ wins this game if the formula is satisfied in this transition system. For simplicity, we here assume that all the bound variables X (in ηX) are distinct in a formula.

Definition 2.2 (Evaluation games). *Given a sentence of modal μ-calculus φ and a transition system $\mathbb{S} = (S, R, \mathcal{I})$, we define the evaluation game $\mathcal{E}(\varphi, \mathbb{S})$ with players $\exists$ and $\forall$ moving a token along positions of the form (ψ, s), where ψ is a subformula of φ and $s \in S$.*

As shown in the following table, the form of the subformula in a position determines the turn of the two players and also their admissible moves for a next position. Player $\exists$'s purpose is to show φ is satisfied at s, while player $\forall$'s goal is opposite.

Positions for player $\exists$	Admissible moves for player $\exists$
$(\psi_1 \vee \psi_2, s)$	$\{(\psi_1, s), (\psi_2, s)\}$
$(\Diamond\psi, s)$	$\{(\psi, t) \mid (s, t) \in R\}$
$(\bot, s)$	$\emptyset$
(P, s) and $s \notin \mathcal{I}(P)$	$\emptyset$
$(\neg P, s)$ and $s \in \mathcal{I}(P)$	$\emptyset$
$(\mu X.\psi_X, s)$	$\{(\psi_X, s)\}$
(X, s) for some subformula $\mu X.\psi_X$	$\{(\psi_X, s)\}$
Positions for player $\forall$	Admissible moves for player $\forall$
$(\psi_1 \wedge \psi_2, s)$	$\{(\psi_1, s), (\psi_2, s)\}$
$(\Box\psi, s)$	$\{(\psi, t) \mid (s, t) \in R\}$
$(\top, s)$	$\emptyset$
(P, s) and $s \in \mathcal{I}(P)$	$\emptyset$
$(\neg P, s)$ and $s \notin \mathcal{I}(P)$	$\emptyset$
$(\nu X.\psi_X, s)$	$\{(\psi_X, s)\}$
(X, s) for some subformula $\nu X.\psi_X$	$\{(\psi_X, s)\}$

In an evaluation game $\mathcal{E}(\varphi, \mathbb{S})$ with an initial position $(\varphi, s_{\mathsf{in}})$, the two players can produce a sequence of positions obeying the above rules as follows,

$$\rho = (\varphi_0, s_0)(\varphi_1, s_1)(\varphi_2, s_2) \ldots \text{ with } (\varphi_0, s_0) = (\varphi, s_{\mathsf{in}})$$

which is called a *play* in the evaluation game $\mathcal{E}(\varphi, \mathbb{S})$.

Note that a play can be finite or infinite. For an infinite play ρ, $\mathsf{Inf}(\rho)$ denotes the set of bound variables X of ηX in φ that are unfolded infinitely often during ρ.

If ρ is finite, player $\exists$ ($\forall$) wins with ρ if the play ends with the position of player $\forall$ ($\exists$). If ρ is infinite, player $\exists$ ($\forall$) wins with ρ if the outmost ηX with $X \in \mathsf{Inf}(\rho)$ is μX (νX). Such ρ's are called *winning plays* for the player who wins with ρ.

A position (φ, s) is said to be a *winning position* for player $\exists$ if player $\exists$ can always produce a winning play starting from (φ, s) no matter how his opponent player $\forall$ plays. By $\mathsf{Win}^{\exists}(\mathcal{E}(\varphi, \mathbb{S}))$, we denote the *winning region*, *i.e.* the set of winning positions, of player $\exists$ in game $\mathcal{E}(\varphi, \mathbb{S})$.

The game semantics of the modal μ-calculus is formulated as follows.

Definition 2.3 (Game semantics of modal μ-calculus). *Let φ be an L_μ-sentence, $\mathbb{S} = (S, R, \mathcal{I})$ be a transition system and $s \in S$. Then φ is said to be game-theoretically satisfied at s, denoted as $(\mathbb{S}, s) \models \varphi$, if $(\varphi, s) \in \text{Win}^\exists(\mathcal{E}(\varphi, \mathbb{S}))$.*

One can show that denotational semantics and game semantics are equivalent, cf. Grädel *et al.*[17] and Venema.[33] So there is no problem in using the expression $(\mathbb{S}, s) \models \varphi$. Also note that with a proper coloring function, the evaluation game turns out to be a parity game which we will introduce later.

Example 2.1. Consider an L_μ-formula $\varphi = \mu X.\Box X$. A play in the evaluation game of φ is

$$\rho = (\mu X.\Box X, s_0)(\Box X, s_0)(X, s_1)(\Box X, s_1)(X, s_2)\cdots$$

At this time, if the play ρ is infinite, then $\forall$ wins as outmost ηX with $X \in \text{Inf}(\rho)$ is μX. On the other hand, if the play is finite, it must be the case that $\forall$ can not have a move from some $(\Box X, s_n)$, therefore $\exists$ wins. That is, φ only holds for a pointed transition system such that there is no infinite path.

2.2. *Alternation hierarchy, variable hierarchy and weak hierarchy*

2.2.1. *Alternation hierarchy of modal μ-calculus*

There are three different alternation hierarchies in term of ways of counting alternations of μ and ν in a formula. Among others, we first introduce the Emerson-Lei hierarchy.[16]

Definition 2.4 (Alternation hierarchy of L_μ). *The (Emerson-Lei) alternation hierarchy of modal μ-calculus is defined as follows.*

- Σ_0^μ, Π_0^μ: *the class of formulas with no fixpoint operators*

- Σ_{n+1}^μ: is the least class of formulas containing $\Sigma_n^\mu \cup \Pi_n^\mu$ and closed under the operations $\vee, \wedge, \square, \lozenge, \mu X$ and the substitution: for a $\varphi(X) \in \Sigma_{n+1}^\mu$ and a closed $\psi \in \Sigma_{n+1}^\mu$, $\varphi(X \backslash \psi) \in \Sigma_{n+1}^\mu$.
- Dually for Π_{n+1}^μ
- $\Delta_n^\mu := \Sigma_n^\mu \cap \Pi_n^\mu$

A formula is strict Σ_n^μ if it is in $\Sigma_n^\mu \setminus \Pi_n^\mu$.

The other two hierarchies are defined as follows. The *simple hierarchy* $\Sigma_n^{S\mu}$ can be obtained from Definition 2.4 by omitting the substitution. The *Niwiński hierarchy* $\Sigma_n^{N\mu}$ is constructed from Definition 2.4 by replacing the condition of the substitution "ψ is closed" with "no free variable of ψ is captured by φ". Obviously, we have

Proposition 2.1. *For all* n, $\Sigma_n^{S\mu} \subsetneq \Sigma_n^\mu \subsetneq \Sigma_n^{N\mu}$.

So far, the alternation hierarchies are defined in a syntactic way. In the following, by the alternation hierarchy of modal μ-calculus, we always mean the semantical counterpart in a certain set $\mathbb{TR}$ of transition systems. The semantical hierarchy is defined as follows:

- $\Sigma_n^\mu[\mathbb{TR}] = \{ [\![\varphi]\!]^\mathbb{S} \mid \varphi \in \Sigma_n^\mu \wedge \mathbb{S} \in \mathbb{TR} \}$,
- $\Pi_n^\mu[\mathbb{TR}] = \{ [\![\varphi]\!]^\mathbb{S} \mid \varphi \in \Pi_n^\mu \wedge \mathbb{S} \in \mathbb{TR} \}$,
- $\Delta_n^\mu[\mathbb{TR}] = \Sigma_n^\mu[\mathbb{TR}] \cap \Pi_n^\mu[\mathbb{TR}]$.

The strictness of alternation hierarchy of modal μ-calculus was first established by Bradfield over the *recursively presented transition systems*.[10–12] A recursively presentable transition system (for short, *rpts*) is a transition system of the form $(S, E, \mathcal{I})$, where each of S and E can be recursively coded as a set of integers and $\mathcal{I}$ is recursive. For simplicity, S can be considered to be a recursive set of natural numbers. Around the same time, Lenzi[21] also proved the strictness over n-ary trees.

Bradfield[10] also indicated that certain formulas expressing winning positions in parity games with n colors, which were introduced by Emerson and Jutla[15] and Walukiewicz,[34] are strict formulas of level n in the alternation hierarchy of modal μ-calculus. The winning region of player $\exists$ in the parity game with n colors can be expressed by the following L_μ-formula

$$\mathcal{W}_n = \eta X_n \cdots \mu X_1 \nu X_0 \left(\bigvee_{i \leq n} (P \wedge P_i' \wedge \Diamond X_i) \vee \bigvee_{i \leq n} (\neg P \wedge P_i' \wedge \Box X_i) \right),$$

where P denotes a position of player $\exists$'s turn, P_i' a position assigned the priority i, $\eta = \nu$ if n is even and $\eta = \mu$ if n is odd.

Moreover, it is known that the alternation hierarchy is strict over some classes of transition systems, for instance, over reflexive transition systems,[1] reflexive and symmetric transition systems.[13] On the other hand, the hierarchy collapses with respect to some classes of transition systems, like finite directed acyclic transition systems.[25]

2.2.2. Variable hierarchy of modal μ-calculus

An analogue measure for modal μ-calculus, defined by the number of distinct bound variables appearing in a formula, induces the *variable hierarchy* of modal μ-calculus. For any n, $L_\mu[n]$ denotes the set of L_μ-formulas that have at most n distinct bound variables, and likewise for $\Sigma_i^\mu[n]$, $\Pi_i^\mu[n]$, $\Delta_i^\mu[n]$ for all level i. For instance, we can define the alternation hierarchy of $L_\mu[1]$ by modifying the definition of alternation hierarchy for L_μ, via level-by-level restricting the formulas with only one fixpoint variable in Definition 2.4, e.g., the semantical $\Sigma_n^\mu[1] = \{ [\![\varphi]\!]^{\mathbb{S}} \mid \varphi \in \Sigma_n^\mu \cap L_\mu[1], \mathbb{S} \in \mathbb{TR} \}$.

It is known that formulas expressing the winning regions of parity games exhaust the finite levels of alternation hierarchy of L_μ. Berwanger[4] further showed that when the number of bound variable is taken into consideration, all such formulas can be reduced to $L_\mu[2]$.

Theorem 2.1. *The alternation hierarchy of $L_\mu[2]$ is strict and not contained in any Σ_n^μ for any $n \in \omega$.*

That is, formulas in $L_\mu[2]$ can express properties in any level of alternation hierarchy of L_μ. Then, it is natural to ask whether $L_\mu[2]$ is as expressive as the whole modal μ-calculus. This problem was negatively answered by Berwanger, Grädel and Lenzi[5,6] by showing the strictness of variable hierarchy as follows:

Theorem 2.2. *For any n, there exists a formula $\varphi \in L_\mu[n]$ which is not equivalent to any formula in $L_\mu[n-1]$.*

2.2.3. Weak hierarchy of modal μ-calculus

Several versions of weak hierarchies are already known.[23] Our version is obtained by generalizing the one-variable alternation hierarchy.

Definition 2.5 (Weak alternation hierarchy of L_μ). *The weak alternation hierarchy of modal μ-calculus is defined as follows.*

- $\Sigma_0^{W\mu}, \Pi_0^{W\mu}$: *the class of formulas with no fixpoint operators*
- $\Sigma_{n+1}^{W\mu}$: *is the least class of formulas containing $\Sigma_n^{W\mu} \cup \Pi_n^{W\mu}$ and closed under the operations $\vee, \wedge, \Box, \Diamond$ and the substitution: for a $\varphi(X) \in \Sigma_1^\mu$ and a closed $\psi \in \Sigma_{n+1}^{W\mu}$, $\varphi(X\backslash\psi) \in \Sigma_{n+1}^{W\mu}$.*
- *Dually for $\Pi_{n+1}^{W\mu}$*
- $\Delta_n^{W\mu} := \Sigma_n^{W\mu} \cap \Pi_n^{W\mu}$

A formula is strict $\Sigma_n^{W\mu}$ if it is in $\Sigma_n^{W\mu} \backslash \Pi_n^{W\mu}$.

Note that in the substitution, $\varphi(X)$ is a usual Σ_1^μ-formula, and also that $\Sigma_{n+1}^{W\mu}$ is not closed under μX. Thus, once an evaluation game enters the scope of some μ (or ν), it doesn't move out no matter how we play. In other words, the alternating layers of fixpoint operators are almost independent from one another. In particular, the one-variable alternation hierarchy also captures this feature well. By definition, we have

Proposition 2.2. *For all n, $\Sigma_n^\mu[1] \subsetneq \Sigma_n^{W\mu} \subsetneq \Delta_2^\mu$.*

Modal μ-formulas in the weak alternation hierarchy are called *weak modal μ-formulas*.

Example 2.2. The following formula φ_1 is purely a one-variable L_μ-formula ($\Pi_2^\mu[1]$). For readability, it may be rewritten as φ_2, a one-variable formula in a broad sense.

- $\varphi_1 = \nu X.\Box(\mu X.\Diamond X) \vee X$.
- $\varphi_2 = \nu X.\Box(\mu Y.\Diamond Y) \vee X$.

And, the following formula φ_3 is a weak modal μ-formula (in fact $\Pi_2^{W\mu}$), but not one-variable.

- $\varphi_3 = \nu X.\Box\nu Z.((\mu Y.\Diamond Y) \wedge \Box X) \vee Z$.

We define the semantical weak alternation hierarchy as we defined the semantical alternation hierarchy of L_μ:

- $\Sigma_n^{W\mu}[\mathbb{TR}] = \{[\![\varphi]\!]^{\mathbb{S}} \mid \varphi \in \Sigma_n^{W\mu} \wedge \mathbb{S} \in \mathbb{TR}\}$,
- $\Pi_n^{W\mu}[\mathbb{TR}] = \{[\![\varphi]\!]^{\mathbb{S}} \mid \varphi \in \Pi_n^{W\mu} \wedge \mathbb{S} \in \mathbb{TR}\}$,
- $\Delta_n^{W\mu}[\mathbb{TR}] = \Sigma_n^{TR\mu}[\mathbb{TR}] \cap \Pi_n^{W\mu}[\mathbb{TR}]$.

3. Hierarchies of one-variable modal μ-calculus and weak parity games

In this section, we discuss relations between the weak and one-variable alternation hierarchies, and weak parity games.

3.1. *Parity games and weak parity games*

A *parity game* $\mathcal{G} = (V_\exists, V_\forall, E, \Omega)$ with index n is played on a colored directed graph, where each node is colored by the priority function $\Omega : V_\exists \cup V_\forall \to \{0, \ldots, n\}$. Two players, $\exists$ and $\forall$, move a token along the edges of the graph, which results in a path, called a *play*. For any position $v \in V_\exists \cup V_\forall$, if $v \in V_\exists$ ($V_\forall$), $\exists$ ($\forall$) chooses a successor v' such that $(v, v') \in E$. The winner of a finite play is the player whose opponent is unable to move. The winner of an infinite play is determined by the priorities appearing in the play. Typically, player $\exists$ wins an infinite play if the largest priority that occurs infinitely often in the play is even, $\forall$ wins otherwise. Such a determination condition is called a *parity condition*. We denote by $\mathrm{Win}^\exists(\mathcal{G})$ the set of winning positions for player $\exists$. For details on parity games, refer to Grädel *et al.*[17]

A parity game $\mathcal{G} = (V_\exists, V_\forall, E, \Omega)$ is said to be *weak* if the coloring function Ω has the following additional property:

for all $v, v' \in V_\exists \cup V_\forall$, if $(v, v') \in E$, then $\Omega(v') \leq \Omega(v)$.

We can think the evaluation game of a weak modal μ-formula as a weak parity game. In fact, given a pointed transition systems $(\mathbb{S}, s_0)$ and a weak modal μ-formula φ, we define a weak parity game $\mathcal{G}$ on a tree, which is equivalent to the evaluation game $\mathcal{E}$ of $(\mathbb{S}, s_0) \models \varphi$.

The arena of $\mathcal{G}$ is defined to be a tree constructed as follows:

(1) each node ρ is a partial play (*i.e.*, a finite sequence of admissible moves) of the evaluation game $\mathcal{E}$; the ownership of each node is inherited from the evaluation game,

(2) the relation of the arena is inherited from the admissible moves in the evaluation game $\mathcal{E}$.

The coloring function Ω of game $\mathcal{G}$ for a weak modal μ-formula φ is defined by cases mainly on the last element $\mathsf{LAST}(\rho)$ of a partial play ρ in $\mathcal{G}$.

- if $\mathsf{LAST}(\rho) = (\phi, s)$ and ϕ has no variables, then $\Omega(\rho) := 0$.
- If $\mathsf{LAST}(\rho) = (\phi, t)$ and $(\nu X.\phi, s)$ appears in ρ with $\nu X.\phi \in \Pi_1^{\mathrm{W}\mu}$, then $\Omega(\rho) := 0$.
- if $\mathsf{LAST}(\rho) = (\phi, t)$, $(\mu X.\psi, s)$ appears in ρ with $\mu X.\psi \in \Sigma_{2n+2}^{\mathrm{W}\mu} \setminus \Pi_{2n+1}^{\mathrm{W}\mu}$ and variable X associated with $\mu X.\psi$ freely occurs in ϕ, then $\Omega(\rho) := 2n + 1$.
- if $\mathsf{LAST}(\rho) = (\phi, t)$, $(\nu X.\psi, s)$ appears in ρ with $\nu X.\psi \in \Pi_{2n+3}^{\mathrm{W}\mu} \setminus \Sigma_{2n+2}^{\mathrm{W}\mu}$ and variable X associated with $\nu X.\psi$ freely occurs in ϕ, then $\Omega(\rho) := 2n + 2$.
- if all the formulas in a partial play ρ (starting with φ) are not of the form $\eta X.\psi$, then

$$\Omega(\rho) := \begin{cases} 2n + 1, & \text{if } \varphi \in \Sigma_{2n+2}^{\mathrm{W}\mu} \setminus \Pi_{2n+1}^{\mathrm{W}\mu}, \\ 2n + 2, & \text{if } \varphi \in \Pi_{2n+3}^{\mathrm{W}\mu} \setminus \Sigma_{2n+2}^{\mathrm{W}\mu}. \end{cases}$$

It is not difficult to see that the coloring function Ω is well-defined, since once an evaluation game of a weak modal μ-formula enters the scope of some μ (or ν), it doesn't move out no matter how we play (cf. Definition 2.5).

3.2. *Strictness of one-variable and weak alternation hierarchies*

The next theorem was first proved by Mostowski.[27] We here present an easier method to show the strictness of weak alternation hierarchy, hence also for the one-variable alternation hierarchy. Our approach is based on the work of Alberucci in Grädel *et al.*,[17] but Alberucci worked with alternating tree automata and here we work directly with the weak modal μ-calculus.

Theorem 3.1. *The semantic weak alternation hierarchy is strict.*

Inspired by Bradfield,[10] we recursively define a weak modal μ-formula $\mathcal{W}'_n$ (in fact, one-variable formula) to express the winning region of player $\exists$ for weak parity games as follows: If P denotes a position of player $\exists$'s turn, and P'_i a position with priority i, then

- $\mathcal{W}'_0 = \nu X.(P \wedge P'_0 \wedge \Diamond X) \vee (\neg P \wedge P'_0 \wedge \Box X)$,
- $\mathcal{W}'_{n+1} = \eta X.(P \wedge P'_{n+1} \wedge \Diamond X) \vee (\neg P \wedge P'_{n+1} \wedge \Box X) \vee \mathcal{W}'_n$ for $n \geq 0$,

where η is μ if n is even, otherwise ν. Notice that $\mathcal{W}'_{2n}$ is a $\Pi^{\mu}_{2n+1}[1]$-formula, and $\mathcal{W}'_{2n+1}$ is a $\Sigma^{\mu}_{2n+2}[1]$-formula.

In the following claims, we first show that $\mathcal{W}'_n$ indeed describes the winning positions for $\exists$ in a weak parity game with colors up to n, and then they witness the strictness of the one-variable alternation hierarchy.

Claim 3.1. *Fix $n \geq 0$. Let $\mathcal{G} = (V_\exists, V_\forall, E, \Omega)$ be a weak parity game with $\Omega(V_\exists \cup V_\forall) = \{0, \dots, n\}$, $\mathbb{S} = (S, R, \mathcal{I})$ be a transition system and $P'_0, \dots, P'_n, P \in \mathfrak{P}$ such that the following holds:*

- $S = V_\exists \cup V_\forall$,
- $R = E$,
- $\mathcal{I}(P'_i) = \Omega^{-1}(\{i\})$ *for $n = 0, \cdots n$, and $\mathcal{I}(P) = V_\exists$,*
- $\mathcal{I}(Q) = \emptyset$ *if $Q \notin \{P'_0, \dots, P'_n, P\}$.*

Then, the following holds:

$$\forall v_0 \in S, \quad (\mathbb{S}, v_0) \models \mathcal{W}'_n \iff v_0 \in \text{Win}^\exists(\mathcal{G}).$$

Proof. Suppose $(\mathbb{S}, v_0) \models \mathcal{W}'_n$ and let σ be a winning strategy for $\exists$ in the evaluation game. We describe a winning strategy σ' for $\mathcal{G}$ using σ. Let s be a vertex with priority i owned by $\exists$, we compute $\sigma(\Diamond X_i, s) = (X_i, t)$ and put $\sigma'(s) = t$. Given an infinite play ρ starting from v_0 and played according to σ', there is j which is eventually the only j occurring in ρ. This means that eventually only vertices with priority j occur on the play, and when we look back at the corresponding play ρ' in the evaluation game according to σ, X_j is eventually the only variable which occurs. As σ is a winning strategy for the evaluation game, X_j must be bound by ν, and so j must be even by the definition of $\mathcal{W}_n''$. Therefore $\exists$ wins with ρ and v_0 is a winning position for $\exists$ in $\mathcal{G}$.

If v_0 is a winning position for $\exists$ in $\mathcal{G}$, let σ be a winning strategy for $\exists$. We define a winning strategy σ' for $\exists$ in the evaluation game of $(\mathbb{S}, v_0) \models \mathcal{W}_n'$. On a node of the form $\Big((P \wedge P_i' \wedge \Diamond X) \vee (\neg P \wedge P_i' \wedge \Box X) \vee \mathcal{W}_{i-1}', s \Big)$, $\exists$ plays:

- $\Big((P \wedge P_i' \wedge \Diamond X), s \Big)$ if $\Omega(s) = i$ and $s \in V_\exists$,
- $\Big((\neg P \wedge P_i' \wedge \Box X), s \Big)$ if $\Omega(s) = i$ and $s \in V_\forall$, and
- $\Big(\mathcal{W}_{i-1}', s \Big)$ if $\Omega(s) < 1$.

For a node in the form of $\Big(\Diamond X, s \Big)$, $\exists$ plays $\Big(X, \sigma(s) \Big)$. We can show σ' is a winning strategy for the evaluation game similar to how we did in the last paragraph. $\qquad \square$

Let φ be a L_μ-formula. Given a pointed transition system $(\mathbb{S}, s_0)$, we can define the evaluation game $\mathcal{E}$ of $(\mathbb{S}, s_0)$ associated with φ. We then transform $\mathcal{E}$ into a parity game $\mathcal{G}$, and then transform $\mathcal{G}$ into a transition system $\mathbb{S}'$.

We define $f_\varphi(\mathbb{S}, s_0) = (\mathbb{S}', (\varphi, s_0))$. That is, f_φ maps a pointed transition system to its evaluation game relative to φ (as a transition system).

Let $(\mathbb{S}, s_0), (\mathbb{T}, t_0)$ be pointed transition systems without loops in their graphs. $(\mathbb{S}, s_0)$ is *isomorphic* to $(\mathbb{T}, t_0)$ if and only if there is a bijection $I : S \to T$ such that:

- $I(s_0) = t_0$,
- for all $s_1, s_2 \in S$, $s_1 E s_2$ iff $I(s_1) E I(s_2)$, and
- for all $s \in S$, for all proposition symbol P, P holds on s iff P holds on $I(s)$.

For all natural number $n \geq 1$, let $(\mathbb{S} \restriction n, s_0)$ be a sub-transition-system that restricts $\mathbb{S}$ to the elements which have distance less than n from s_0. We say $(\mathbb{S}, s_0)$ is *n-isomorphic* to $(\mathbb{T}, t_0)$ if and only if $(\mathbb{S} \restriction n, s_0)$ is isomorphic to $(\mathbb{T} \restriction n, t_0)$.

Fix some L_μ-formula φ. If $(\mathbb{S}, s_0)$ and $(\mathbb{T}, t_0)$ are n-isomorphic via a function I, $f_{\varphi \wedge \varphi}(\mathbb{S}, s_0)$ and $f_{\varphi \wedge \varphi}(\mathbb{T}, t_0)$ are n-isomorphic via a function J defined by

$$J(\psi, s) = (\psi, I(s)) \qquad \text{for } s \in S \text{ and } \psi \text{ a subformula of } \varphi.$$

We now have the tools to show the next lemma.

Claim 3.2. *For all $\varphi \in \Sigma_n^{W\mu}$, $f_{\varphi \wedge \varphi}$ has a fixed point, i.e., there is a transition system $(\mathbb{T}, t)$ such that $f_\varphi(\mathbb{T}, t)$ is isomorphic to $(\mathbb{T}, t)$.*

Proof. Let $(\mathbb{S}, s_0)$ be any given pointed transition system. $f_{\varphi \wedge \varphi}$ is also a transition system, and on $(\varphi \wedge \varphi, s_0)$ only P and a fixed P_i hold. Note that i depends solely on the choice of φ.

We define $(\mathbb{T}_0, t_0)$ to be a pointed transition system with only one node t_0 on its graph, and only P and the P_i fixed above hold. From the above paragraph, we have that $(\mathbb{T}_0, t_0)$ and $f_{\varphi \wedge \varphi}(\mathbb{T}_0, t_0)$ are 1-isomorphic.

For $n \in \omega$, let us define $(\mathbb{T}_{n+1}, t_{n+1}) = f_{\varphi \wedge \varphi}(\mathbb{T}_n, t_n)$. We already have that $(\mathbb{T}_0, t_0)$ and $(\mathbb{T}_1, t_1)$ are 1-isomorphic, and we can show that $(\mathbb{T}_n, t_n)$ and $(\mathbb{T}_{n+1}, t_{n+1})$ are $(n+1)$-isomorphic by induction on n.

We can show that, for all $n \geq 1$, $(\mathbb{T}_n, t_n)$ is n-isomorphic to all $(\mathbb{T}_m, t_m)$ with $m > n$. Therefore we can define a transition system $(\mathbb{T}, t)$ which is n-isomorphic to $(\mathbb{T}_n, t_n)$ for all n. To define $\mathbb{T}$, we may take $(\mathbb{T}_n \upharpoonright n, t_n)$ and $(\mathbb{T}_{n+1} \upharpoonright n, t_{n+1})$ to be equal, as they are n-isomorphic, so the graph of $\mathbb{T}$ is the union of the graph of the $\mathbb{T}_n \upharpoonright n$, the valuation of $\mathbb{T}$ is the union of the valuation of the $\mathbb{T}_n \upharpoonright n$ and t is t_1. Finally, we have that $(\mathbb{T}, t)$ is isomorphic to $f_{\varphi \wedge \varphi}(\mathbb{T}, t)$ as they are n-isomorphic for all n. □

We can now prove Theorem 3.1.

Proof of Theorem 3.1. For simplicity, let n be even. The case for n odd is symmetric. We prove by contradiction. We know that $\mathcal{W}_n' \in \Pi_{n+1}^{W\mu}$, suppose that $\mathcal{W}_n'$ is equivalent to some formula in $\Pi_n^{W\mu}$. Let $\varphi \in \Sigma_n^{W\mu}$ be equivalent to $\neg \mathcal{W}_n'$.

Let $(\mathbb{T}, t_0)$ be a fixed point of $f_{\varphi \wedge \varphi}$. Then

$$
\begin{aligned}
(\mathbb{T}, t_0) \models \neg \mathcal{W}_n' &\iff (\mathbb{T}, t_0) \models \varphi \wedge \varphi \\
&\iff f_{\varphi \wedge \varphi}(\mathbb{T}, t_0) \models \mathcal{W}_n' \\
&\iff (\mathbb{T}, t_0) \models \mathcal{W}_n'
\end{aligned}
$$

which reaches a contradiction. $\qquad\square$

The proof of Theorem 3.1 gives us:

Theorem 3.2. *The semantic alternation hierarchy for one-variable L_μ-formulas is strict.*

Proof. For simplicity, let n be even. We also know that $\mathcal{W}_n' \in \Pi_{n+1}^\mu[1]$. By way of contradiction, we had assumed that $\mathcal{W}_n' \in \Pi_n^\mu[1]$. Then, this would imply $\mathcal{W}_n' \in \Pi_n^{\mathrm{W}\mu}$, since every one-variable formula is a weak formula by Proposition 2.2. Hence, the proof of Theorem 3.1 would lead to a contradiction. This completes the proof. $\qquad\square$

Now let $\mathbb{TR}'$ be the class of recursively presentable transition systems (*rpts*). Instead of considering f_φ, we consider a function f_φ' that takes an *rpts* $(\mathbb{S}, s_0)$ to a normalized form of the transition system $f_\varphi(\mathbb{S}, s_0)$.

Let φ be a fixed formula of the form $\psi \wedge \psi$ and $\mathbb{S} = (S, R, \mathcal{I})$ be an *rpts*. We suppose that all the bound variables in φ are distinct. Suppose that we have a Gödel numbering of the formulas. Define $f_\varphi'(\mathbb{S}, s_0) = (S', R', \mathcal{I}')$ with initial point s_0' such that

$$
\chi_{S'}(s) = 1 \iff s = 0 \vee s = \langle a, b, c, d \rangle
$$
$$
\text{where } a \text{ is a subformula of } \varphi, \; b \text{ is a state of } S,
$$
$$
\text{and } d \in \{0, 1\}
$$

$$\langle a, b \rangle \in R' \iff \chi_{S'}(a) = \chi_{S'}(b) = 1 \wedge$$
$$[(a = 0 \wedge b = \langle \psi, s_0, 0, j \rangle \text{ with } j \in \{0, 1\})$$
$$\vee (a = \langle \psi_0 \circ \psi_1, s, i, j \rangle \wedge b = \langle \psi_k, s, i, j \rangle$$
$$\text{with } k \in \{0, 1\})$$
$$\vee (a = \langle \triangle\psi, s, i, j \rangle \wedge b = \langle \psi, t, i, j \rangle \wedge R(s, t))$$
$$\vee (a = \langle \eta X.\psi, s, i, j \rangle \wedge b = \langle \psi, s, i, j \rangle)$$
$$\vee (a = \langle X, s, i, j \rangle \wedge b = \langle \psi_X, s, i+1, j \rangle)]$$

(In the above definition $\triangle$ is $\square$ or $\lozenge$ and $\circ$ is $\wedge$ or $\vee$.) We define $\mathcal{I}'(P_i)$ recursively using S' and $\mathcal{I}$. We can also assign the ownership of the vertices using P. The initial state of the new *rpts* is $\langle \phi, 0 \rangle$. This transition system is isomorphic to $f_\varphi(\mathbb{S}, s_0)$ and so we get the following claim as a corollary of the proof of Claim 3.2.

Claim 3.3. *For all $\varphi \in \Sigma_n^{W\mu}$, $f'_{\varphi \wedge \varphi}$ has a fixed point in $\mathbb{TR}'$; i.e., there is a transition system $(\mathbb{T}, t) \in \mathbb{TR}'$ such that $f'_\varphi(\mathbb{T}, t)$ is isomorphic to $(\mathbb{T}, t)$.*

The proof of Theorem 3.1 also works with $\mathbb{TR}'$ and f'_φ. Indeed, we can modify the definition of f'_φ to define directly the fixed point of $f_{W_n \wedge W_n}$ by exchanging χ_S and R by $\chi_{S'}$ and R' in the definition above. So given an *rpts* $(\mathbb{S}, s_0)$, $f_\varphi(\mathbb{S}, s_0)$ is also an *rpts*. Therefore:

Corollary 3.1. *The semantic weak alternation hierarchy restricted to recursively presentable transition systems is strict.*

Corollary 3.2. *The semantic alternation hierarchy for one-variable L_μ-formulas restricted to recursively presentable transition systems is strict.*

4. Transfinite extensions

4.1. Transfinite μ-calculus

Rabin's classical result[29] implies that Δ_2^μ is equal to the compositions of Σ_1^μ and Π_1^μ, or essentially equal to the class of weak formulas in the view of tree automata. However, we can show that in term of parity games, Δ_2^μ is properly stronger than the weak modal μ-formulas.

Corollary 4.1. *There are formulas in Σ_2^μ and Π_2^μ which define the winning region of any weak parity game, but there is no weak modal μ-formulas for which the same holds.*

Proof. By our proof of the strictness on the weak alternation hierarchy, there is no formula equivalent to $\mathcal{W}_n'$ in $\Sigma_n^{W\mu} \cup \Pi_n^{W\mu}$, so the claim about weak modal μ-formulas hold.

For any weak parity game $\mathcal{G}$, we regard it as a non-weak parity game $\mathcal{G}'$ with priorities 0 (in place of even priorities in $\mathcal{G}$) and 1 (in place of odd priorities in $\mathcal{G}$), then the winning regions of weak parity games can be defined by

$$\mu X_0.\nu X_1.(P \wedge P_0' \wedge \Diamond X_0) \vee (P \wedge P_1' \wedge \Diamond X_1) \vee (\neg P \wedge P_0' \wedge \Box X_0) \vee (\neg P \wedge P_1' \wedge \Box X_1),$$

and

$$\nu X_1.\mu X_0.(P \wedge P_0' \wedge \Diamond X_0) \vee (P \wedge P_1' \wedge \Diamond X_1) \vee (\neg P \wedge P_0' \wedge \Box X_0) \vee (\neg P \wedge P_1' \wedge \Box X_1).$$

$\square$

Now, we want to extend one-variable formulas to infinite length to get more expressive power. In fact, Bradfield et al.[7] already introduced a transfinite version of modal μ-calculus, and investigated its relation to the Gale-Stewart games in the transfinite difference hierarchy over Σ_2^0 sets. We here consider a transfinite extension of weak modal μ-calculus in a similar way. From now, we are allowed to use infinite disjunction $\bigvee_i \Phi_i$ and conjunction $\bigwedge_i \Phi_i$ for countably (recursively) many formula Φ_i. Note that we require that finitely many variables occur in $\bigwedge_i$, and for our results we may even require that only one variable occurs in our formulas. We define the transfinite hierarchy of the L_μ-formulas as follows:

Definition 4.1 (Transfinite hierarchy of L_μ). *For a successor ordinal $\alpha + 1$, a $\Sigma_{\alpha+1}^\mu$ ($\Pi_{\alpha+1}^\mu$) formula is constructed from Σ_α^μ and Π_α^μ in a similar way as when α is finite. For a limit ordinal λ, a Σ_λ^μ (Π_λ^μ) is an infinitary formula in the form $\bigvee_i \Phi_i$ ($\bigwedge_i \Phi_i$) where Φ_i is Σ_α^μ with $\alpha < \lambda$.*

We define the transfinite hierarchy of one-variable formulas as in section 2, that is, $\Sigma_\alpha^\mu[1] = \Sigma_\alpha^\mu \cap L_\mu[1]$. We also define the transfinite hierarchy of weak modal μ-formulas $\Sigma_\alpha^{W\mu}(\Pi_\alpha^{W\mu})$ in the same manner as Definition 4.1.

We can extend the weak parity games by allowing infinite priorities with priority functions of the form $\Omega : V_\exists \cup V_\forall \to \alpha$, where α is some ordinal number. Let $\mathcal{G}$ be a weak parity game using priorities up to ω. The winning region for this game can be expressed by the following $\Sigma_{\omega+1}^{W\mu}$ one-variable formula

$$\mathcal{W}_\omega := \nu X. \left[\left(P \wedge P_\omega \wedge \Diamond X \right) \vee \left(\neg P \wedge P_\omega \wedge \Box X \right) \vee \bigvee_{n<\omega} W_n \right].$$

Note, however, any transfinite weak parity game can be expressed by a two-priority parity game by regarding even (odd) ordinals as 0 (1), as explained in Corollary 4.1.

In general, we define the formulas that express the winning regions of weak parity games allowing infinite many priorities as follows:

- $\mathcal{W}_0' := \nu x.[(P \wedge P_0 \wedge \Diamond X) \vee (\neg P \wedge P_0 \wedge \Box X)].$
- $\mathcal{W}_{\alpha+1}' := \eta X.[(P \wedge P_{\alpha+1} \wedge \Diamond X) \vee (\neg P \wedge P_{\alpha+1} \wedge \Box X) \vee \mathcal{W}_\alpha]$ where $\eta = \nu$ if $\alpha + 1$ is even and $\eta = \mu$ if $\alpha + 1$ is odd.
- $\mathcal{W}_\lambda' := \nu X.[(P \wedge P_\lambda \wedge \Diamond X) \vee (\neg P \wedge P_\lambda \wedge \Box X) \vee \bigvee_{\alpha<\lambda} \mathcal{W}_\alpha']$ where λ is a limit ordinal and assumed to be even.

As in the finite case, $\mathcal{W}_\alpha'$ is a $\Pi_{\alpha+1}^\mu[1]$-formula if α is even, and $\mathcal{W}_\alpha'$ is a $\Sigma_{\alpha+1}^\mu[1]$-formula if α is odd. The following theorem holds by essentially the same proof as Theorem 3.1.

Theorem 4.1. *The transfinite alternation hierarchy for one-variable L_μ-formulas is strict.*

At last, we note that the one-variable μ-calculus can define formulas which are not equivalent to any formula of the finite modal μ-calculus. The following formula φ is in $\Sigma_\omega^\mu[1]$:

$$\varphi = \bigwedge_{i\in\omega} \Diamond^i \left[\left(\bigwedge_{j<i} \neg P_j \right) \wedge P_i \right]$$

Here, $\Diamond^i$ is a sequence of i many $\Diamond$s. We can show that φ is satisfied by some (recursively presented) transition system and that all models of φ are infinite. As the finite modal μ-calculus has the Finite Model Property,[18] φ is not in Σ_n^μ for any $n \in \omega$.

5. Transfinite weak μ-arithmetic

In this section we explore a connection between the transfinite weak alternation hierarchy and weak parity games derived from the transfinite difference hierarchy of Σ_1^0. To this end, we start with weak μ-arithmetic. We let $\mathbb{TR}'$ be the set of all $rpts$'s.

The weak μ-arithmetic is obtained by adding the fixed point operators μ and ν to the first-order arithmetic. In this context, $\mu x X.\varphi$ is the least fixed point of the operator $\Gamma : \mathcal{P}(\omega) \to \mathcal{P}(\omega)$ defined by $\Gamma(X) = \{x \in \omega : \varphi(x, X)\}$, and $\nu x X.\varphi$ is the greatest fixed point of the same operator. Note that μ and ν are dual by $\neg[\mu x X.\varphi(X)] = \nu x X.\neg\varphi(\neg X)$.

Example 5.1. The following term defines the even numbers in the μ-arithmetic:

$$\mu x X.(x = 0 \vee (x - 2) \in X).$$

We also recall the difference hierarchy for Σ_2^0 as follows:

Definition 5.1. For each $\alpha < \omega_1^{\mathrm{ck}}$,

$$S \in \Sigma_\alpha^{\delta,2} \iff S = \bigcup_{\beta \in I} A_\beta \setminus \left(\bigcup_{\zeta < \beta} A_\zeta \right).$$

where $(A_\beta)_{\beta < \alpha}$ is an effective enumeration of a sequence of sets in Σ_2^0 and I is the set of ordinals less that α whose parity is opposite to the parity of α. The limit ordinals are considered to be even.

Note that if we substitute Σ_2^0 by other pointclasses we can get the corresponding difference hierarchies.

Recall that Tanaka[26,31,32] showed that in the Baire space, Δ_{n+1}^0 is exhausted by the difference hierarchy over Σ_n^0 up to ω_1^{ck}. Depending on this theorem for $n = 2$, Bradfield et al.[7] established the correspondence between the transfinite alternation hierarchy and parity games with infinite priorities derived from the transfinite difference hierarchy of Σ_2^0:

Theorem 5.1 (Bradfield, Duparc, Quickert). *For all $\alpha < \omega_1^{ck}$,*
$$\eth\Sigma_\alpha^{\delta,2} = \Sigma_{\alpha+1}^\mu[\mathbb{TR}'].$$

Here $\eth$ is the game quantifier, defined by
$$\eth\xi.P(\alpha,\vec{x}) = \{\vec{x} : \exists \text{ wins the Gale-Steward game with}$$
$$\text{payoff } P(\xi,\vec{x})\} \subseteq \omega^k\},$$
where $P \subseteq \omega^\omega \times \omega^k$ for some $k \in \omega$.

In the same fashion, we will consider the correspondence between the transfinite weak alternation hierarchy and weak parity games with infinite priorities (that is, weak parity games with infinite priorities derived from the transfinite difference hierarchy of Σ_1^0).

Definition 5.2 (Weak alternation hierarchy for the μ-arithmetic). *The weak alternation hierarchy for the μ-arithmetic formulas and set terms are defined as follows:*

- $\Sigma_0^{W\mu}$ *is the set of all the first order formulas and all set variables.*
- $\Sigma_1^{W\mu}$ *is the closure of $\Sigma_0^{W\mu}$ under μ and $\in$.*
- $\Sigma_{\alpha+1}^{W\mu}$ *is generated from $\Sigma_\alpha^{W\mu} \cup \Pi_\alpha^{W\mu}$ by closing it under $\vee, \wedge, \in$ and the following substitution rules:*

   (a) *If $\varphi(X)$ is Σ_1^μ and if ψ is a $\Sigma_{\alpha+1}^{W\mu}$ term without free set variables, then $\varphi(X\backslash\psi)$ is also $\Sigma_{\alpha+1}^{W\mu}$;*

   (b) *if φ is a Σ_1^μ of μ-arithmetic, φ' is a subformula of φ and ψ is a $\Sigma_{\alpha+1}^{W\mu}$ formula without free set variables, then $\varphi(\varphi'\backslash\psi)$ is also $\Sigma_{\alpha+1}^{W\mu}$.*

- *If λ is a limit ordinal, then $\Sigma_\lambda^{W\mu}$ is generated from $\bigcup_{\alpha<\lambda}\Sigma_\alpha^{W\mu}$ and closed under $\bigvee_{i<\omega}$.*
- $\Pi_\alpha^{W\mu}$ *contains all the negations of formulas and set terms in $\Sigma_\alpha^{W\mu}$.*
- $\Pi_\lambda^{W\mu}$ *contains all the negations of $\Sigma_\lambda^{W\mu}$ formulas and terms.*

Observe that we abuse the notation of substitution in this definition. This is necessary in the transfinite levels of the weak hierarchy, as there are no weak μ-term strictly in the limit levels.

Bradfield[11] proved a correspondence between the alternation hierarchies of the modal μ-calculus and μ-arithmetic over *rpts*. We can adapt his proof to show

Theorem 5.2. *Let $\varphi(z)$ be a $\Sigma_\alpha^{W\mu}$ formula of μ-arithmetic. There is a rpts T, a valuation V and a $\Sigma_\alpha^{W\mu}$ formula $\overline{\varphi}$ of L_μ such that $\varphi(s)$ iff $s \in \|\overline{\varphi}\|_V^T$.*

Theorem 5.3. *For each L_μ formula $\varphi \in \Sigma_\alpha^{W\mu}$ and for each recursively presentable transition system T, $\|\varphi\|^T \subseteq \omega$ is definable by a $\Sigma_\alpha^{W\mu}$ formula of μ-arithmetic.*

That is, in the viewpoints of expressive power, the αth level of the alternation hierarchy of L_μ restricted to *rpts*'s is equal to the αth level of the alternation hierarchy of the μ-arithmetic.

Let α be computable and $\Sigma^{\delta,1}$ be the αth level of the difference hierarchy of Σ_1^0. We now show that a set of natural numbers is definable by a $\Sigma_{\alpha+1}^{W\mu}$ formula of μ-arithmetic if and only if it of the form ∂P where P is in $\Sigma_\alpha^{\delta,1}$, that is:

Theorem 5.4. $\Sigma_{\alpha+1}^{W\mu}[\mathbb{TR}'] = \partial\Sigma_\alpha^{\delta,1}$, *for all $\alpha < \omega_1^{\text{ck}}$.*

Proof. We follow the proof of Theorem 5.1, but we modify Gale and Stewart's proof of Open Determinacy.

Claim 5.1. $\partial\Sigma_\alpha^{\delta,1} \subseteq \Sigma_{\alpha+1}^{W\mu}[\mathbb{TR}']$, *for all $\alpha < \omega_1^{\text{ck}}$.*

Proof. We prove this based on Gale and Steward's proof of Open Determinacy.

Before formalizing the proof we sketch its idea. The proof of Open Determinacy proceeds by defining the winning position of the game

$$\exists n.Q(\xi[n], \vec{x}),$$

where Q is a recursive set and $\vec{x}$ are natural number parameters. We omit the parameters most of the time. We define the winning positions inductively, starting with the "easy" positions:

$$s \in W_0 \text{ iff } Q(s).$$

If we have defined W_α, we define $W_{\alpha+1}$ by adding to W_α all the positions where $\exists$ can immediately force the play into W_α, *i.e.*,

$s \in W_{\alpha+1}$ iff $s \in W_\alpha$

$\qquad \lor \; lh(s)$ is odd and $\forall n.s \frown n \in W_\alpha$

$\qquad \lor \; lh(s)$ is even and $\exists m \forall n.s \frown m \frown n \in W_\alpha$,

where $\frown$ denotes sequence concatenation. For a limit ordinal λ, we define $W_\lambda = \bigcup_{\alpha < \lambda} W_\alpha$.

We can extend this proof to the difference hierarchy for Σ_1^0 by adding a $\Pi_\alpha^{\delta,1}$ parameter $R(\xi)$ to the open game. This results in the game

$$\exists n.Q(\xi[n]) \land R(\xi).$$

If we denote the winning region of the game with payoff R by W_R, we define the winning region of the new game by:

$s \in W_0$ iff $Q(s) \land s \in W_R$, and

$s \in W_{\alpha+1}$ iff $s \in W_\alpha$

$\qquad \lor \; lh(s)$ is odd and $\forall n.s \frown n \in W_\alpha \cap W_R$

$\qquad \lor \; lh(s)$ is even and $\exists m \forall n.s \frown m \frown n \in W_\alpha \cap W_R$.

Let $W = \bigcup_\alpha W_\alpha$, we can then show that

- There is a winning strategy for player $\exists$ iff $\langle\rangle \in W$.
- There is a winning strategy for player $\forall$ iff $\langle\rangle \notin W$.

We now formalize the above argument in the μ-arithmetic by defining the winning region with μ-arithmetic formulas.

- Let $\alpha = 0$. We show that $\partial \Sigma_0^{\delta,1} \subseteq \Sigma_1^{W\mu}$. Consider the game $P(\xi, \vec{x}) = \exists n.Q(\xi(n))$ where Q has no unbounded quantifiers. We define its winning positions by:

$$W_P = \mu y Y.\Big(Q(y)$$

$$\lor \; (lh(y) \text{ is odd and } \forall n.s \frown n \in Y)$$

$$\lor \; (lh(y) \text{ is even and } \exists m \forall n.s \frown m \frown n \in Y)\Big).$$

We have then that $\vec{x} \in \partial \xi.P(\xi, \vec{x})$ iff $\langle\rangle \in W(\vec{x})$.

- We show that $\partial \Sigma_{\alpha+1}^{\delta,1} \subseteq \Sigma_{\alpha+2}^{W\mu}$. Consider the game $P(\xi, \vec{x}) = \exists n.Q(\xi(n)) \wedge R(\xi)$, where Q is recursive and R is $\Pi_\alpha^{\delta,1}$. We define its winning positions by:

$$W_P = \mu y Y. \Big((Q(y) \wedge W_R(y))$$

$$\vee \ (lh(y) \text{ is odd and } \forall n.y \frown n \in Y)$$

$$\vee \ (lh(y) \text{ is even and } \exists m \forall n.y \frown m \frown n \in Y) \Big).$$

We have then that $\vec{x} \in \partial \xi. P(\xi, \vec{x})$ iff $\langle \rangle \in W(\vec{x})$ again.

- Let λ be a limit ordinal and suppose $\partial \Sigma_\alpha^{\delta,1} \subseteq \Sigma_{\alpha+1}^{W\mu}$ holds for all $\alpha < \lambda$. Let $P(\xi, \vec{x}) = \bigvee_{i<\omega} \varphi_n$ be a $\Sigma_\lambda^{\delta,1}$ game, with each φ_i in $\bigcup_{\alpha<\lambda} \Sigma_\lambda^{\delta,1}$. Then

$$W_P = \mu y Y. \Big(\bigvee_{n<\omega} W_{\varphi_n}(y)$$

$$\vee \ (lh(y) \text{ is odd and } \forall n.y \frown n \in Y)$$

$$\vee \ (lh(y) \text{ is even and } \exists m \forall n.y \frown m \frown n \in Y) \Big).$$

Yet another time, $\vec{x} \in \partial \xi. P(\xi, \vec{x})$ iff $\langle \rangle \in W(\vec{x})$.

We can then conclude that $\partial \Sigma_\alpha^{\delta,1} \subseteq \Sigma_{\alpha+1}^{W\mu}$ for all $\alpha < \omega_1^{ck}$. □

Claim 5.2. *Every weak formula and term of μ-arithmetic can be put in a normal form as follows:*

- *If $\varphi \in \Sigma_1^{W\mu}$ is a term, then $\varphi \equiv \mu x_1 X_1.\psi$ where ψ is an arithmetical formula.*
- *If $\varphi \in \Sigma_{\alpha+1}^{W\mu}$ is a term, then $\varphi \equiv \mu x_{\alpha+1} X_{\alpha+1}.\psi(\psi_1, \ldots, \psi_n)$ where ψ is an arithmetical formula and $\psi_1, \ldots, \psi_n$ are $\Pi_\alpha^{W\mu}$ terms in normal form.*
- *If α is not a limit ordinal and $\varphi \in \Sigma_\alpha^{W\mu}$ is a formula, then $\varphi \equiv \tau_\alpha \in \varphi'$ where $\varphi' \in \Sigma_\alpha^{W\mu}$ is a term in normal form.*
- *If λ is a limit ordinal and $\varphi \in \Sigma_\lambda^{W\mu}$ is a formula, then $\varphi \equiv \bigvee_{n\in\omega} \psi_n$ where each ψ_n is a formula in $\bigcup_{\alpha<\lambda} \Sigma_\alpha^{W\mu}$. We furthermore suppose that X_β is a μ-variable iff α and β have the same parity.*

Note that if λ is a limit ordinal, then there are no μ-terms in $\Sigma_\lambda^{W\mu} \setminus \bigcup_{\alpha < \lambda} \Sigma_\alpha^{W\mu}$.

Proof. Use Lubarsky's Normal Form Theorem for μ-arithmetic[24] and observe that using the available arithmetical machinery we can do all the substitutions at the same time. We can resolve the weak parity condition on the limit case by padding the terms with μ and ν operators. $\qquad\square$

Claim 5.3. $\Sigma_{\alpha+1}^{W\mu}[\mathbb{TR}'] \subseteq \partial\Sigma_\alpha^{\delta,1}$, *for all* $\alpha < \omega_1^{ck}$.

Proof. Let $\varphi(\vec{x})$ denote a $\Sigma_{\alpha+1}^{W\mu}$ term of μ-arithmetic in normal form. The key fact is that the model checking game for a $\Sigma_{\alpha+1}^{W\mu}$-formula is essentially the code of a $\Sigma_\alpha^{\delta,1}$ set. In case of $\alpha = 0$, if $\exists$ wins the model checking game G for $n \in \varphi(\vec{x})$ he wins in finite time. So G is an open game. For other cases, we use a more sophisticated argument which holds for the successor and limit cases.

Suppose there is some $n \in \varphi$. Consider the model checking game for $n \in \varphi$. We think of this game as a tree and increasingly simplify it. First, we can code the game as a subtree of $\omega^{<\omega}$. Furthermore, we can suppose that this tree has no finite maximal branches. To simplify the tree, we suppose each node of the tree marks a loopback, *i.e.*, that each node is of the form $n' \in X_\beta$. We also omit the n's and keep track only of the βs, so each vertice is just an ordinal number β. We denote the tree obtained by this process by T.

For each successor $\beta \in \alpha$, we define

$$C_\beta = \left\{ x \in [T] \mid \exists n.x(n) = \beta \right\}.$$

Each C_β is in Σ_1^0. Remember that an ordinal is odd (even) iff it is of the form $\omega\lambda + n$ where $n \in \omega$ is odd (even). Let $I = \{\beta < \alpha : parity(\beta) \neq parity(\alpha)\}$. Define

$$C = \bigcup_{\beta \in I} C_\beta \setminus \left(\bigcup_{\zeta < \beta} C_\zeta \right)$$

C is a $\Sigma_\alpha^{\delta,1}$ set. We show C is the winning set for $\exists$.

Let $x \in C$ and fix the β such that $x \in C_\beta \setminus \left(\bigcup_{\zeta<\beta} C_\zeta\right)$. We have that the β with parity not equal to the parity of α are all ν-variables, by the definition of the normal form. Therefore x is a play that eventually loops the ν-variable X_β. That is, x is a winning play for $\exists$.

If x is a play that is winning for $\exists$, then it eventually inspects only some ν-variable X_β, so $x \in C_\beta \setminus \left(\bigcup_{\zeta<\beta} C_\zeta\right)$ and consequently $x \in C$. $\qquad\qquad\qquad\qquad\qquad\qquad\qquad\qquad\qquad\qquad\qquad\qquad$ □

This concludes the proof of Theorem 5.4.

Tanaka[32] showed that $\bigcup_{\alpha<\omega_1^{ck}} \Sigma_\alpha^{\delta,1} = \Delta_2^0$. As a consequence we have that $\bigcup_{\alpha<\omega_1^{ck}} \Sigma_\alpha^{W\mu}[\mathbb{TR}'] = \partial\Delta_2^0$ holds. By Theorem 5.1, we have that $\partial\Delta_2^0 = \Delta_2^\mu[\mathbb{TR}']$, so it follows:

Corollary 5.1. $\bigcup_{\alpha<\omega_1^{ck}} \Sigma_\alpha^{W\mu}[\mathbb{TR}'] = \Delta_2^\mu[\mathbb{TR}']$.

We conjecture that given a natural number n, $\bigcup_{\alpha<\omega_1^{ck}} \Sigma_\alpha^{nW\mu}[\mathbb{TR}'] = \Delta_{n+1}^\mu[\mathbb{TR}']$, where $\Sigma_\alpha^{nW\mu}$ is defined in the same manner as $\Sigma_\alpha^{W\mu}$ but replacing the occurrences of Σ_1^μ by Σ_n^μ.

6. Conclusion

In this paper, we considered one-variable modal μ-calculus and its relations to weak modal μ-calculus and weak parity games. Forming translations among them such that their alternation depths and parities are associated to each other, we proved the strictness of these hierarchies. Furthermore, we introduced their transfinite extensions and established the correspondence between the transfinite weak alternation hierarchy and weak parity games with infinite priorities derived from the transfinite difference hierarchy.

The one-variable μ-calculus is a very useful tool to make transfinite models easier to build and handle. In our future work, we will apply one-variable and weak model techniques to investigate the properties of the upper level of hierarchy.

When we consider infinite priorities on infinite game graphs, "recursively presented" transition systems may be too general. Probably, it is more natural to treat, for instance, pushdown game graphs, and then the stack heights and stack contents are another important parameters.

References

1. L. Alberucci and A. Facchini, The modal μ-calculus hierarchy over restricted classes of transition systems. *J. Symb. Log.* **74** (2009), 1367-1400.
2. A. Arnold and D. Niwinski, *Rudiments of μ-Calculus.* Studies in Logic and the foundations of Mathematics, volume 146, North-Holland, Amsterdam (2001).
3. A. Arnold, The μ-calculus alternation-depth hierarchy is strict on binary trees. *RAIRO-Theor. Inf. Appl.* **33** (1999), 329-339.
4. D. Berwanger, Game logic is strong enough for parity games. *Studia Logica* **75** (2003), 205-219.
5. D. Berwanger and G. Lenzi, The variable hierarchy of the μ-calculus is strict, in *22th Annual Symposium on Theoretical Aspects of Computer Science, STACS 2005. Lecture Notes in Comput. Sci.* **3404** (2005), 97-109.
6. D. Berwanger, E. Grädel and G. Lenzi, The variable hierarchy of the μ-calculus is strict. *Theory Comput. Syst.* **40** (2007), 437-466.
7. J.C. Bradfield, J. Duparc and S. Quickert, Transfinite extension of the mu-calculus, in *Proceedings of 19th International Workshop Computer Science Logic, CSL'2005. Lecture Notes in Comput. Sci.* **3634** (2005), 384-396.
8. J.C. Bradfield, Fixpoints, games and the difference hierarchy. *RAIRO-Theor. Inf. Appl.* **37** (2003), 1-15.
9. J.C. Bradfield, Fixpoint alternation: arithmetic, transition systems and the binary tree. *RAIRO-Theor. Inf. Appl.* **33** (1999), 341-356.
10. J.C. Bradfield, Simplifying the modal mu-calculus alternation hierarchy, in *Proceedings of 15th Annual Symposium on Theoretical Aspects of Computer Science, STACS 1998. Lecture Notes in Comput. Sci.* **1373** (1998), 39-49.
11. J.C. Bradfield, The modal μ-calculus alternation hierarchy is strict. *Theoret. Comput. Sci.* **195** (1998), 133-153.
12. J.C. Bradfield, The modal μ-calculus alternation hierarchy is strict, in *Proceedings of the 7th International Conference on Concurrency Theory, CONCUR 1996. Lecture Notes in Comput. Sci.* **1119** (1996), 233-246.
13. G. D'Agostino and G. Lenzi, On modal μ-calculus over reflexive symmetric graphs. *J. Logic Comput.* **23** (2013), 445-455.

14. G. D'Agostino and G. Lenzi, On the μ-calculus over transitive and finite transitive frames. *Theoret. Comput. Sci.* **411** (2010), 4273-4290.

15. E.A. Emerson and C.S. Jutla, Tree automata, mu-calculus and determinacy, in *Proc. FOCS' 91. 32nd Annual Symposium on. IEEE* IEEE Computer Society Press (1991), 368-377.

16. E.A. Emerson and C.L. Lei, Efficient model checking in fragments of the propositional mu-calculus, in *Proc. First IEEE Symp. on Logic in Computer Science* (1986), pp. 267-278.

17. E. Grädel, W. Thomas, T. Wilke (Eds.), Automata Logics, and Infinite Games. in: LNCS, vol. 2500, Springer, 2002.

18. D. Kozen, A finite model theorem for the propositional μ-calculus. *Studia Logica* **47.3** (1988), 233-241.

19. D. Kozen, Results on the propositional μ-calculus. *Theoret. Comput. Sci.* **27** (1983), 333-354.

20. K. Lehtinen *Syntactic complexity in the modal μ-calculus.* Ph.D. thesis. University of Edinburgh, Edinburgh (2017).

21. G. Lenzi, A hierarchy theorem for the μ-calculus, in *Proceedings of the 25th International Colloquium on Automata, Languages and Programming, ICALP 1996.* **1099** *Lecture Notes in Comput. Sci.* (1996), 87-97.

22. G. Lenzi, A hierarchy theorem for the mu-calculus, Proceedings of the 23rd International Colloquium on Automata, Languages and Programming, ICALP'96, LNCS, vol. 1099, Springer-Verlag, 1996, pp. 87-97.

23. K. Lehtinen, Syntactic complexity in the modal μ calculus, Doctoral thesis, the University of Edinburgh, 2017.

24. R.S. Lubarsky, μ-definable sets of integers. *J. Symb. Log.* **58** (1993), 291-313.

25. R. Mateescu, Local model-checking of modal mu-calculus on acyclic labeled transition systems, in *TACAS* Springer Berlin Heidelberg (2002), 281-295.

26. M.Y.O. MedSalem and K. Tanaka, Δ_3^0-determinacy, comprehension and induction. *J. Symb. Log.* **72** (2007) 452-462.

27. A.W. Mostowski, Hierarchies of weak automata and weak monadic formulas. *Theoret. Comput. Sci.* 83 (1991), 323-335.

28. D. Niwiński, Fixed point characterization of infinite behavior of finite-state systems. *Theoret. Conput. Sci.* 189 (1997), 1-69.

29. M.O. Rabin, Weakly definable relations and special automata. *Stud. Logic Found. Math.* **59** (1970) 1-23.

30. L. Santocanale and A. Arnold, Ambiguous classes in μ-calculi hierarchies. *Theoret. Comput. Sci.* **333** (2005), 265-296.

31. K. Tanaka, *Descriptive set theory and subsystems of analysis.* Ph.D. thesis. University of California, Berkeley (1986).

32. K. Tanaka, Weak axioms of determinacy and subsystems of analysis I: Δ_2^0-games, *Zeitschrift für mathematische Logik und Grundlagen der Mathematik,* vol. 36 (1990), 481-491.
33. Y. Venema, Lectures on the modal μ-calculus. 2008.
34. I. Walukiewicz, Monadic second-order logic on tree-like structures. *Theoret. Comput. Sci.* **271** (2002), 311-346.
35. T. Wilke, Alternating tree automata, parity games, and modal μ-calculus. *Bull. Soc. Math. Belg.* **8** (2001), 359-391.

Infinite Games, Inductive Definitions and Transfinite Recursion

Kazuyuki Tanaka

Mathematical Institute, Tohoku University,
Sendai, Miyagi 9808578, Japan
E-mail: tanaka.math@tohoku.ac.jp

Keisuke Yoshii

National Institute of Technology, Okinawa College,
Nago, Okinawa 9052192, Japan
E-mail: kyoshii@okinawa-ct.ac.jp

The purpose of this research is to investigate the logical strength of determinacy of Gale-Stewart games from the standpoint of reverse mathematics. It has been known for a few decades that the determinacy of Σ_1^0 sets (open sets) is equivalent to system ATR_0 and that of Σ_2^0 corresponds to the axiom of Σ_1^1 inductive definitions. Recently, much effort has been made to characterize the determinacy of game classes above Σ_2^0 within second order arithmetic.

In this paper, we introduce an axiom of transfinite recursion of inductive definitions, denoted $[\Sigma_1^1]^k$-IDTR, to pin down the determinacy of $\Delta((\Sigma_2^0)_{k+1})$ games. Here, $(\Sigma_2^0)_{k+1}$ is the difference class of $k+1$ Σ_2^0 sets and $\Delta((\Sigma_2^0)_{k+1})$ is the conjunction of $(\Sigma_2^0)_{k+1}$ and co-$(\Sigma_2^0)_{k+1}$. By $[\Sigma_1^1]^k$-ID, we denote inductive definition with k $[\Sigma_1^1]$ operators.

Keywords: Gale-Stewart Games; Reverse Mathematics; Inductive Definitions; Transfinite Recursion.

1. Introduction

Initial researches on the determinacy of Gale-Stewart games have been conducted mainly in descriptive set theory.[7] It was provable in ZFC that a Borel game is determinate, but the same thing does not hold for an analytic game. These facts simply represent that the strength of the determinacy of games varies depending on the complexity of their winning sets.

Subsequently, researches classifying the strength of determinacy within second order arithmetic have been started along the Reverse Mathematics program. This foundational program aims for answering the following questions: *What set existence axioms are needed to prove the theorems of ordinary mathematics?* See Simpson[11] for the major results.

To begin with, Steel[12] established the equivalence between Σ_1^0 (open) determinacy and ATR_0 (see also Tanaka[13]), and then Tanaka[13] showed that $\Sigma_1^0 \wedge \Pi_1^0$ determinacy and $\Pi_1^1\text{-}\mathsf{CA}_0$ are equivalent, and moreover introduced a new axiom $\Pi_1^1\text{-}\mathsf{TR}_0$ to pin down Δ_2^0 determinacy. To characterize Σ_2^0 (F_σ) determinacy, Tanaka[14] formulated the axiom of Σ_1^1 inductive definitions in second order arithmetic and proved their equivalence. At the level of Δ_3^0 ($F_{\sigma\delta} \cap G_{\delta\sigma}$), Med-Salem and Tanaka[6] proved that Δ_3^0 determinacy is equivalent to an axiom of transfinite combinations of Σ_1^1 inductive definitions (over $\Pi_3^1\text{-}\mathsf{TI}_0$). Subsequently, Welch[16] proved that $\Pi_3^1\text{-}\mathsf{CA}_0$ implies Π_3^0 determinacy while $\Delta_3^1\text{-}\mathsf{CA}_0$ does not. Finally, Montalbán and Shore[8] showed that for any $m \geq 1$, $\Pi_{m+2}^1\text{-}\mathsf{CA}_0$ proves the determinacy of $(\Sigma_3^0)_m$ sets, but $\Delta_{m+2}^1\text{-}\mathsf{CA}_0$ does not. Here, $(\Sigma_n^0)_m = \Sigma_n^0 \wedge \neg(\Sigma_n^0)_{m-1}$ (See Definition 3.2). Thus, $(\Sigma_3^0)_{<\omega}$ determinacy is not provable over Z_2. Conversely, it is known[6] that even Δ_1^1 determinacy does not imply Δ_2^1 comprehension.

In this paper, we focus on the characterization of finer hierarchies of $(\Sigma_2^0)_{<\omega}$ determinacy in terms of inductive definitions and transfinite recursion. More precisely, we introduce a system $[\Sigma_1^1]^k\text{-}\mathsf{IDTR}_0$, which guarantees the transfinite recursion of inductive definitions with multiple Σ_1^1 operators and show the following equivalence(Theorem 6.6): over ACA_0, for any $k > 0$, the following are equivalent,

(1) $\Delta((\Sigma_2^0)_{k+1})\text{-}\mathsf{Det}$.

(2) $\mathsf{Sep}(\Delta_2^0, (\Sigma_2^0)_k)\text{-}\mathsf{Det}$.

(3) $[\Sigma_1^1]^k\text{-}\mathsf{IDTR}_0$

An important feature characterizing the present paper is a unified method to prove equivalence between inductive definitions and corresponding determinacy. In order to do so, we repeatedly nest a previous argument within a new construction. We also use the equivalence of $\mathsf{Sep}(\Delta_n^0, (\Sigma_n^0)_k)$ and $\Delta((\Sigma_n^0)_{k+1})$ for any $k \geq 1$. See Definition 3.4 for details.

Heinatsch and Möllerfeld[3] proved that $\Pi_2^1\text{-}\mathsf{CA}_0$ is conservative over $(\Sigma_2^0)_{<\omega}$ determinacy with respect to the Π_1^1-sentences. Thus, finer studies on $(\Sigma_2^0)_{<\omega}$ determinacy help shed light on properties of $\Pi_2^1\text{-}\mathsf{CA}_0$ and other relevant systems.

The following diagram shows the results on determinacy strength of Δ_3^0 games in second order arithmetic. The left column contains subsystems of second order arithmetic from weaker to stronger. The right column contains classes of the games in the Baire space. Each row represents that a certain axiom is equivalent to the determinacy of the corresponding games over appropriate systems (ACA_0, but with $\Pi_3^1\text{-}\mathsf{TI}$ for the last row).

Subsystem of SOA	Determinacy in Baire space
ATR_0	Δ_1^0 Σ_1^0
$\Pi_1^1\text{-}\mathsf{CA}_0$	$\Delta((\Sigma_1^0)_2) = \mathsf{Sep}(\Delta_1^0, \Sigma_1^0)$ $(\Sigma_1^0)_2$
$\Pi_1^1\text{-}\mathsf{TR}_0$	Δ_2^0
$\Sigma_1^1\text{-}\mathsf{ID}_0$	Σ_2^0 $\mathsf{Sep}(\Delta_1^0, \Sigma_2^0)$ $\mathsf{Sep}(\Sigma_1^0, \Sigma_2^0)$
$\Sigma_1^1\text{-}\mathsf{IDTR}_0$	$\Delta((\Sigma_2^0)_2) = \mathsf{Sep}(\Delta_2^0, \Sigma_2^0)$
$\vdots$	$\vdots$
$[\Sigma_1^1]^k\text{-}\mathsf{ID}_0$	$(\Sigma_2^0)_k$
$[\Sigma_1^1]^k\text{-}\mathsf{IDTR}_0$	$\Delta((\Sigma_2^0)_{k+1}) = \mathsf{Sep}(\Delta_2^0, (\Sigma_2^0)_k)$
$\vdots$	$\vdots$
$[\Sigma_1^1]^{\mathrm{TR}}\text{-}\mathsf{ID}_0$	Δ_3^0

Additionally, Nemoto, MedSalem and Tanaka[9] studied determinacy strength of infinite games in the Cantor space and compared them with these of infinite games in the Baire space.

2. Preliminaries

In this section, we recall some basic definitions and facts about second order arithmetic. The language $\mathcal{L}_2$ of second-order arithmetic is a two-sorted language consisting of constant symbols $0, 1, +, \cdot,$ $=, <$ with number variables $x, y, z, \ldots$ and unary function variables $f, g, h, \ldots$. We also use set variables $X, Y, Z, \ldots$, intending to range over the $\{0, 1\}$-valued functions, that is, the characteristic functions of sets.

The formulas can be classified as follows:

- φ is *bounded* (Π_0^0) if it is built up from atomic formulas by using propositional connectives and bounded number quantifiers ($\forall x < t$), ($\exists x < t$), where t does not contain x.
- φ is Π_0^1 if it does not contain any function quantifier. Π_0^1-formulas are called *arithmetical* formulas.
- $\neg\varphi$ is Σ_n^i if φ is a Π_n^i-formula ($i \in \{0, 1\}, n \in \omega$).
- $\forall x_1 \cdots \forall x_k \varphi$ is Π_{n+1}^0 if φ is a Σ_n^0-formula ($n \in \omega$),
- $\forall f_1 \cdots \forall f_k \varphi$ is Π_{n+1}^1 if φ is a Σ_n^1-formula ($n \in \omega$).

We loosely say that a formula is Σ_n^i (resp. Π_n^i) if it is equivalent over a base theory (such as ACA_0) to a $\psi \in \Sigma_n^i$ (resp. Π_n^i).

We now define some popular axiom schemata of second order arithmetic.

Definition 2.1. Let $\mathcal{C}$ be a set of $\mathcal{L}_2$-formulas.

(1) $\mathcal{C}$-IND: $(\varphi(0) \wedge \forall x(\varphi(x) \to \varphi(x+1))) \to \forall x \varphi(x)$,
   where $\varphi(x)$ belongs to $\mathcal{C}$.
(2) $\mathcal{C}$-TI: for any well-ordering $<_X$, $(\forall x(\forall y <_X x \; \varphi(y) \to \varphi(x))) \to \forall x \varphi(x)$,
   where $\varphi(x)$ belongs to $\mathcal{C}$.
(3) $\mathcal{C}$-CA: $\exists X \forall x(x \in X \leftrightarrow \varphi(x))$,
   where $\varphi(x)$ belongs to $\mathcal{C}$ and X does not occur freely in $\varphi(x)$.
(4) $\mathcal{C} \cap \mathcal{C}^-$-CA: $\forall x(\varphi(x) \leftrightarrow \psi(x)) \to \exists X \forall x(x \in X \leftrightarrow \varphi(x))$,
   where $\varphi(x)$ and $\neg\psi(x)$ belong to $\mathcal{C}$ and X does not occur freely in $\varphi(x)$.

(5) $\mathcal{C}$-AC : $\forall x \exists X \varphi(x, X) \rightarrow \exists X \forall x \varphi(x, X_x)$,

where $\varphi(x, X)$ belongs to $\mathcal{C}$ and $X_x = \{y : (x, y) \in X\}$.

The system ACA_0 consists of the ordered semiring axioms for $(\omega, +, \cdot, 0, 1, <)$, Σ_1^0-CA and Σ_1^0-IND. For a set Λ of sentences, Λ_0 denotes the system consisting of ACA_0 plus Λ.

By Δ_n^i-CA, we denote $\Sigma_n^i \cap (\Sigma_n^i)^-$-CA. We can easily show that for any $k \geq 0$,

$$\Delta_k^1\text{-}\mathsf{CA}_0 \subset \Sigma_k^1\text{-}\mathsf{AC}_0.$$

Moreover, if $k = 2$, the above two axioms are known to be equivalent to each other.

Finally, we introduce an axiom of determinacy. For a formula φ with a distinct variable f ranging over $\mathbb{N}^{\mathbb{N}}$, we associate a two-person *game* G_φ (or simply denote φ) as follows: player I and player II alternately choose a natural number (starting with player I) to form an infinite sequence $f \in \mathbb{N}^{\mathbb{N}}$ and player I (resp. II) wins iff $\varphi(f)$ (resp. $\neg\varphi(f)$). We say that φ is *determinate* if one of the players has a *winning strategy* $\sigma : \mathbb{N}^{<\mathbb{N}} \rightarrow \mathbb{N}$ in the game φ. For a class $\mathcal{C}$ of formulas, $\mathcal{C}$-Det is the axiom which states that any game in $\mathcal{C}$ is determinate.

3. Difference Sets

Definition 3.1. Let $\mathcal{C}$ and $\mathcal{C}'$ be classes of formulas. We denote the classes of formulas in the form $\varphi \wedge \psi$ ($\varphi \in \mathcal{C}, \psi \in \mathcal{C}'$) as $\mathcal{C} \wedge \mathcal{C}'$, and $\neg\psi$ ($\psi \in \mathcal{C}$) as $\neg\mathcal{C}$.

Definition 3.2. For all $n, k \geq 1$, we define the classes of the formulas $(\Sigma_n^0)_k, (\Pi_n^0)_k$ as follows.

- $(\Sigma_n^0)_1 = \Sigma_n^0$, $\quad (\Sigma_n^0)_k = \Sigma_n^0 \wedge (\Pi_n^0)_{k-1}$ if $k > 1$,
- $(\Pi_n^0)_k = \neg(\Sigma_n^0)_k$.

Remark: $(\Sigma_n^0)_{2k} = (\Sigma_n^0)_2 \vee (\Sigma_n^0)_2 \vee \cdots \vee (\Sigma_n^0)_2$ (k times).

Comment. Lemma 3.1. *For any* $n, k \geq 1$, *the following hold.*

- $(\Sigma_n^0)_k = \Pi_n^0 \wedge (\Sigma_n^0)_{k-1}$ *if* k *is even.*
- $(\Sigma_n^0)_k = \Sigma_n^0 \vee (\Sigma_n^0)_{k-1}$ *if* k *is odd.*

Definition 3.3. Let $\mathcal{C}$ be a class of formulas. A $\mathcal{C}$-formula φ is called a $\Delta(\mathcal{C})$-formula if there exists a $\neg\mathcal{C}$-formula φ' such that φ and φ' are equivalent over an appropriate system, e.g., ACA_0. In particular, we write Δ_n^0 for $\Delta(\Sigma_n^0)$.

Our objective of this paper is to pin down the strength of $\Delta((\Sigma_2^0)_k)$-determinacy. For this purpose, we have the following definition.

Definition 3.4. Let $\mathcal{C}, \mathcal{C}'$ be classes of formulas. A formula $\varphi(f)$ is called a $\mathsf{Sep}(\mathcal{C}, \mathcal{C}')$-formula, if it is written as $(\psi(f) \wedge \eta(f)) \vee (\neg\psi(f) \wedge \eta'(f))$ for some $\varphi \in \mathcal{C}, \eta \in \neg\mathcal{C}$ and $\eta' \in \mathcal{C}'$.

Theorem 3.1. *A $\Delta((\Sigma_n^0)_{k+1})$-formula is equivalent to a $\mathsf{Sep}(\Delta_n^0, (\Sigma_n^0)_k)$-formula for $n, k \geq 1$, and vice versa.*

We first prove $\mathsf{Sep}(\Delta_n^0, (\Sigma_n^0)_k) \subseteq \Delta((\Sigma_n^0)_{k+1})$.

Lemma 3.2. *Suppose $1 \leq k, n(< \omega)$. For any Δ_n^0-formula $\psi(f)$, $(\Sigma_n^0)_k$-formula $\eta(f)$, and $(\Pi_n^0)_k$-formula $\eta'(f)$, there exists $\Delta((\Sigma_n^0)_{k+1})$-formula $\zeta(f)$ such that*
$$\forall f(\zeta(f) \leftrightarrow ((\psi(f) \wedge \eta(f)) \vee (\neg\psi(f) \wedge \eta'(f)))).$$

Proof. Take a Δ_n^0-formula ψ, a $(\Sigma_n^0)_k$-formula η and a $(\Pi_n^0)_k$-formula η'. Then we show by induction on k that $\theta \equiv (\psi \wedge \eta) \vee (\neg\psi \wedge \eta')$ is $\Delta((\Sigma_n^0)_{k+1})$.

First suppose $k = 1$. Then obviously, $(\psi \wedge \eta) \vee (\neg\psi \wedge \eta')$ is $\Sigma_n^0 \vee \Pi_n^0$, i.e., $(\Sigma_2^0)_2$. Its negation $\neg\theta$ can be written as $(\psi \wedge \neg\eta) \vee (\neg\psi \wedge \neg\eta')$, which is also $(\Sigma_2^0)_2$, and then θ is $\Delta((\Sigma_n^0)_2)$.

For the induction step, assume the claim with less than k. By Definition 3.2, η, η' can be written as $\sigma \wedge \eta_1$, $\pi \vee \eta_1'$ respectively, where $\sigma \in \Sigma_n^0, \pi \in \Pi_n^0$, and $\eta_1 \in (\Pi_n^0)_{k-1}, \eta_1' \in (\Sigma_n^0)_{k-1}$.

By the induction hypothesis, $\theta_1 \equiv (\psi \wedge \eta_1) \vee (\neg\psi \wedge \eta_1')$ is $\Delta((\Sigma_n^0)_k)$. We easily observe that

$$\theta \leftrightarrow ((\psi \wedge \sigma) \vee \neg\psi) \wedge \theta_1 \vee (\neg\psi \wedge \pi).$$

Since $(\psi \wedge \sigma) \vee \neg\psi$ is Σ_n^0 and θ_1 is $(\Sigma_n^0)_k$, $((\psi \wedge \sigma) \vee \neg\psi) \wedge \theta_1$ is still $(\Sigma_n^0)_k$. So, θ is $(\Pi_n^0)_{k+1}$ since $\neg\psi \wedge \pi$ is Π_n^0. Similarly, we can show that $\neg\theta$ is also $(\Pi_n^0)_{k+1}$. Hence, θ is $\Delta((\Sigma_n^0)_{k+1})$, which completes the proof. □

To prove $\mathsf{Sep}(\Delta_n^0, (\Sigma_n^0)_k) \supseteq \Delta((\Sigma_n^0)_{k+1})$, we need the following lemma.

Lemma 3.3. *Suppose $k, n \geq 1$. Two disjoint $(\Pi_n^0)_k$-formulas φ_0, φ_1 are separated by a $\mathsf{Sep}(\Delta_n^0, (\Sigma_n^0)_{k-1})$-formula δ, i.e., $\varphi_0 \to \delta \to \neg\varphi_1$ holds.*

Proof. Suppose $k = 1$. Let $\varphi_i \equiv \forall x \theta_i$ with $\theta_i \in \Sigma_{n-1}^0$. Assume that φ_0 and φ_1 are disjoint. So, $\neg\varphi_0 \vee \neg\varphi_1$ always holds.

Now, let $\delta \equiv \exists x(\neg\theta_1 \wedge \forall y < x\theta_0)$. Then, it is easy to see

$$\neg\delta \leftrightarrow \exists x(\neg\theta_0 \wedge \forall y \leq x\theta_1).$$

Thus, δ is Δ_n^0. Also, it is clear that $\delta \to \neg\varphi_1$ and $\neg\delta \to \neg\varphi_0$. Hence, δ separates φ_0 and φ_1.

Suppose $k > 1$. Let $\varphi_i \equiv \pi_i \vee \psi_i$ with $\pi_i \in \Pi_n^0$ and $\psi_i \in (\Sigma_n^0)_{k-1}$. Assume φ_0 and φ_1 are disjoint. Then π_0 and π_1 are also disjoint, and so by the above argument, there exists $\delta \in \Delta_n^0$ such that $\pi_0 \to \delta \to \neg\pi_1$ holds. Now let $\overline{\delta} \equiv (\delta \wedge \neg\psi_1) \vee \psi_0$, i.e., $(\delta \wedge \neg\psi_1) \vee (\neg\delta \wedge \psi_0)$. So, $\overline{\delta} \in \mathsf{Sep}(\Delta_n^0, (\Sigma_n^0)_{k-1})$. Then, noticing π_i and ψ_{i-1} are disjoint, we have $\pi_0 \to (\delta \wedge \pi_0) \to (\delta \wedge \neg\psi_1) \to (\neg\pi_1 \wedge \neg\psi_1) \equiv \neg\varphi_1$, and so $(\pi_0 \vee \psi_0) \to \overline{\delta} \to (\neg\varphi_1 \vee \psi_0) \to \neg\varphi_1$, since φ_1 and ψ_0 are also disjoint. □

Now, $\mathsf{Sep}(\Delta_n^0, (\Sigma_n^0)_k) \supseteq \Delta((\Sigma_n^0)_{k+1})$ is straightforward.

Lemma 3.4. *Suppose $k, n \geq 1$. For any $\Delta((\Sigma_n^0)_k)$-formula $\zeta(f)$, there exist a Δ_n^0-formula $\psi(f)$, a $(\Sigma_n^0)_k$-formula $\eta(f)$ and a $(\Pi_n^0)_k$-formula $\eta'(f)$ such that*

$$\forall f(\zeta(f) \leftrightarrow ((\psi(f) \wedge \eta(f)) \vee (\neg\psi(f) \wedge \eta'(f)))).$$

Proof. If ζ is $\Delta((\Sigma_n^0)_k)$, then ζ and $\neg\zeta$ are disjoint $(\Sigma_n^0)_k$. So, by the above lemma, there exists a $\mathsf{Sep}(\Delta_n^0, (\Sigma_n^0)_k)$-formula ψ such that $\zeta \to \psi \to \neg\neg\zeta$, i.e., $\zeta \leftrightarrow \psi$. □

Finally, we recall our effective version[13] of the Hausdorff-Kuratowski theorem, which states that a Δ_n^0 set can be spitted into a transfinite difference of Π_{n-1}^0 sets. By using this result, a $\mathsf{Sep}(\Delta_n^0, (\Sigma_n^0)_k)$ set can be treated as a certain combination of $(\Sigma_n^0)_k$ sets, which enables us to prove $\mathsf{Sep}(\Delta_n^0, (\Sigma_n^0)_k)$-$\mathsf{Det}$ by transfinite recursion. This idea will be explained later in Chapter 5.

4. Inductive Definitions

We start by formalizing the axiom of inductive definition. An operator $\Gamma : P(\mathbb{N}) \to P(\mathbb{N})$ belongs to a class $\mathcal{C}$ of formulas iff its graph $\{(x, X) : x \in \Gamma(X)\}$ is defined by a formula in $\mathcal{C}$. Γ is said to be *monotone* iff $\Gamma(X) \subset \Gamma(Y)$ whenever $X \subset Y$. By mon-$\mathcal{C}$, we will denote the class of monotone operators in $\mathcal{C}$.

A relation W is a *pre-ordering* iff it is reflexive, connected and transitive. W is a *pre-well-ordering* iff it is a well-founded preordering. The *field* of W is the set $F = \{x : \exists y \ (x, y) \in W \lor (y, x) \in W\}$. An axiom of inductive definition asserts the existence of a pre-well-ordering constructed by iterative applications of a given operator.

Definition 4.1. Let $\mathcal{C}$ be a set of $\mathcal{L}_2$ formulas. $\mathcal{C}$-ID asserts that for any operator $\Gamma \in \mathcal{C}$, there exists a set $W \subset \mathbb{N} \times \mathbb{N}$ such that

(1) W is a pre-well-ordering on its field F,
(2) $\forall x \in F \ W_x = \Gamma(W_{<x}) \cup W_{<x}$,
(3) $\Gamma(F) \subset F$,

where $W_x = \{y \in F : (y, x) \in W\}$ and $W_{<x} = \{y \in F : (y, x) \in W$ and $(x, y) \notin W\}$.

$$\Gamma(\emptyset) \cup \Gamma(\Gamma(\emptyset)) \dots \quad \overset{\textstyle x}{W_{<x}} \qquad \overset{\textstyle F \supset \Gamma(F)}{\text{(fixd point)}}$$

Fig. 1. A pre-well-ordering constructed by $\mathcal{C}$-ID_0

We write $\mathcal{C}$-MI to denote mon-$\mathcal{C}$-ID. We note that for a monotone operator Γ, the second condition of the above definition can be replaced by

$$\forall x \in F \ W_x = \Gamma(W_{<x}).$$

It is also easy to see that for any class $\mathcal{C}$, $\mathcal{C}$-MI_0 implies $\mathcal{C}$-CA_0. Remark that while Tanaka[14] studied relations between lightface statements (without parameters), we here focus on boldface statements, since the use of parameters makes arguments much easier (and hence the base system weaker) as seen in the following proofs.

Now, we have

Theorem 4.1 (Tanaka,[14] MedSalem-Tanaka[6]). *Over* ACA_0, *the following are equivalent.*

(1) Σ_2^0-Det.
(2) $\mathsf{Sep}(\Delta_1^0, \Sigma_2^0)$-Det.
(3) Σ_1^1-MI.
(4) Σ_1^1-ID.

Proof. (2)$\Rightarrow$(1) and (4)$\Rightarrow$(3) is obvious. For (3)$\Rightarrow$(1), it suffices to formalize Wolfe's original proof of Σ_2^0-determinacy as shown by Tanaka.[14] For (1)$\Rightarrow$(2), one may treat a $\mathsf{Sep}(\Delta_1^0, \Sigma_2^0)$-game as a Δ_1^0-game with an oracle telling which player has a winning strategy for the Σ_2^0 (or Π_2^0)-game starting at each position. More rigorously, the argument goes as follows.

Let $\xi(f)$ be a $\mathsf{Sep}(\Delta_1^0, \Sigma_2^0)$-formula defined by

$$\xi(f) \leftrightarrow ((\psi(f) \wedge \eta(f)) \vee (\neg\psi(f) \wedge \eta'(f))),$$

where ψ is Δ_1^0, η is Σ_2^0 and η' is Π_2^0. First, consider the Σ_2^0-game $\eta(f)$ starting at a position $p \in \mathbb{N}^{<\mathbb{N}}$. By Σ_2^0-Det, we know that one of the players has a winning strategy for such a game. So, we have

$$\neg\exists\sigma\forall\tau\eta(p * \sigma \otimes \tau) \leftrightarrow \exists\tau\forall\sigma\neg\eta(p * \sigma \otimes \tau),$$

which means that $\exists\sigma\forall\tau\eta(p * \sigma \oplus \tau)$ is Δ_2^1. Assuming Δ_2^1-CA_0, the following set exists.

$$P_\eta = \{p \in \mathbb{N}^{<\mathbb{N}} : \exists\sigma\forall\tau\eta(p * \sigma \otimes \tau) \wedge \text{ the length of } p \text{ is even,}$$
$$\vee\forall\tau\exists\sigma\eta(p * \sigma \otimes \tau) \wedge \text{ the length of } p \text{ is odd.}\}$$

Here, P_η may be viewed as the set of player I's winning positions for η, or an oracle telling which player has a winning strategy for η at a given position. Similarly, by using Δ_2^1-CA_0, we can define $P_{\eta'}$ for the set of player I's winning positions for η'.

Now, consider the Δ_1^0-formula $\psi(f)$ as $\exists n \theta(f[n])$ and its negation as $\exists n \theta'(f[n])$ where θ and θ' are Δ_0^0. Then, the original game $\xi(f)$ is equivalent to the following Δ_1^0-game $\xi'(f)$:

$$\exists n(\theta(f[n]) \wedge f[n] \in P_\eta) \vee \exists n(\theta'(f[n]) \wedge f[n] \in P_{\eta'}).$$

Obviously, if player I has a winning strategy for ξ', he has a winning strategy for ξ. If player II has a winning strategy for ξ' and she follows it, then she can reaches a position p such that $\theta(p) \wedge p \notin P_\eta$ or $\theta'(p) \wedge p \notin P_{\eta'}$. By Σ_2^0-Det, if $p \notin P_\eta$ (or $P_{\eta'}$), then player II has a winning strategy for η (or η'). Hence, player II has a winning strategy for ξ.

In sum, we prove, under the hypothesis of Δ_2^1-CA$_0$, that

$$\Sigma_2^0\text{-Det} \to \text{Sep}(\Delta_1^0, \Sigma_2^0)\text{-Det},$$

or strictly with parameter X indicated,

$$\forall X (\Sigma_2^0\text{-Det}(X) \to \text{Sep}(\Delta_1^0, \Sigma_2^0)\text{-Det}(X)).$$

Since the above sentences is Π_3^1, by Π_3^1-conservation of Δ_2^1-CA$_0$ over Π_1^1-CA$_0$ (See Simpson,[11] Theorem IX.4.9), we can deduce it from Π_1^1-CA$_0$. Noticing the boldface Σ_2^0-Det implies Π_1^1-CA$_0$ ([11]), we conclude that (the boldface) Σ_2^0-Det implies (the boldface) $\text{Sep}(\Delta_1^0, \Sigma_2^0)$-Det.

To show (2) $\Rightarrow$ (4), let Γ be a Σ_1^1-operator. The axiom Σ_1^1-ID asserts the existance of pre-well-ordering V with field F such that $V_x = \Gamma(V_{<x}) \cup V_{<x}$ for all $x \in F$, and such that $F \subset \Gamma(F)$. We will construct a $\text{Sep}(\Delta_1^0, \Sigma_2^0)$-game G such that player I has no winning strategy, and such that for any winning strategy τ of player II, V is Π_1^0 in τ, which suffices to get (2)$\Rightarrow$(4).

The outline of game G is as follows. Player I starts the game with playing a number y^*, intending to raise a question whether or not y^* is in the field of V. In reply to this question, player II may either *accept* y^* or *reject* y^*. If II accepts y^*, II is requested to list the $\leq y^*$-segment of V and also to give certain witnesses for his assertions. The role of player I is then to watch his opponent's moves, and point out a possible error in them. If II rejects y^* at the beginning, the roles are reversed. After the initial stage, the player constructing the $(\leq y^*)$-segment of V is called Pro, and the other player Con. Roughly speaking, Pro wins the game iff Con can not prove that Pro makes a false or erroneous assertion.

Before describing details of the Pro/Con subgame, we remark that player I has no winning strategy in G. By way of contradiction, assume that player I had a winning strategy σ. Let y^* be player I's first move called by σ. Since player II may either accept or reject y^*, player I must defend both cases. In the case player II rejects y^*, player I needs to construct V such that y^* is in the field of V. But if such a construction is successful, player II could accept y^* and win the game by constructing the same V. Therefore, player I has no winning strategy in this game, and so by the determinacy of the game, player II must have a winning strategy, say τ. Finally, we let V be the set of pairs (x, y) such that the strategy τ calls for II as Pro to put (x, y) into the list for V at every meaningful position. Then, we can show that V is actually our desired set.

To describe the Pro/Con subgame, we first fix a Σ_1^1 operator Γ. By the normal form theorem, there exists a recursive relation R such that the following holds:

$$x \in \Gamma(X) \leftrightarrow \exists f \forall z R(f[z], x, X[z]),$$

where $f[z]$ and $X[z]$ encode the finite sequence $\langle f(0), f(1), \ldots, f(z-1) \rangle$ and the finite set $X \cap \{0, 1, \ldots, z-1\}$, respectively. Then, Pro constructs a pre-ordering V_{y^*} and witness $\{f_x : x \in \text{field}(V_{y^*})\}$, and Con makes challenges against Pro together with certain witnesses. The Pro/Con subgame proceeds as follows.

Pro	Con
$v(0), f(0)$	
	$c(0), u(0), g(0)$
$v(1), f(1)$	
	$c(1), u(1), g(1)$
$\vdots$	
	$\vdots$

Note that $u, v \in \{0, 1\}^{\mathbb{N}}$, $f, g \in \mathbb{N}^{\mathbb{N}}$ and $c \in \{\{-1\} \cup \mathbb{N}\}^{\mathbb{N}}$. Intuitively, v is a function to construct a pre-ordering $V = \{(n_0, n_1) : v(n) = 1\}$, where n codes (n_0, n_1), and f is a collection of witnesses for V to reduce its logical complexity. In addition, c is used for Con's challenges, and (u, g) is Con's alternative for (v, f).

For every n such that $v(n) = 1$, Pro is required to construct a witness f^n for $n_0 \in \Gamma(V_{<n_1}) \cup V_{<n_1}$, where $f^n(m) = f(n + m)$. Thus, Pro needs to satisfy the following conditions (their conjunction denoted by $(*1)$).

(a) $V = \{(n_0, n_1) : v(n) = 1\}$ is a pre-ordering with y^* in the field.
(b) $\forall n(v(n) = 1 \to \forall z R(f^n[z], n_0, V_{<n_1}[z]) \vee n_0 \in V_{<n_1})$.

Condition $(*1)$ can be written as a Π_1^0-formula.

Con makes a *challenge* to Pro's negative assertion $(v(n) = 0)$ by playing $c(m) = n$ at stage $m \geq n$. In order to show that Con's challenge is proper, Con constructs an initial segment $U^m \subseteq V_{<n_1}$ and witness g^m by using functions u and g so that $g^m(k) = g(k+m)$ for any k, and $U^m = \{k \in \mathbb{N} : u(k + m) = 1\}$. Thus, Con's challenge $c(m) = n$ is "successful" if

$$U^m \text{ is an initial segment of } V_{<n_1} \text{ and } \forall z\, R(g^m[z], n_0, U^m[z]).$$

Then, the above condition can be written as a Π_1^0-formula. Note that there may be no l such that $U^m = V_{<l}$ if $V_{<n_1}$ is not well-founded.

There is a *rule* which Con must follow whenever Con makes a challenge. Put simply, challenges must be done along the order already constructed by Pro in the decreasing way and below y^*. Namely, once Con challenges to $v(n_0, n_1) = 0$, then Con can challenge to $v(m_0, m_1) = 0$ only if $v(m_1, n_1) = 1$ and $v(n_1, m_1) = 0$ are asserted by Pro before m. See the figure below. Con can not challenge to the shaded part, $V_{y^*} \setminus V_{<n_1}$, after the challenge to $v(n_0, n_1) = 0$. We name

$$n_1 \qquad y^*$$

$$V_{<n_1} \qquad\qquad V_{y^*}$$

Fig. 2.   The Rule of Challenge

this condition as $(*2)$, which can be written as a Π_1^0-formula. From this rule, it is clear that if Con makes infinitely many challenges, then Pro's construction contains an infinite descending sequence.

We now define the winning conditions for Pro and Con.

(a) The case that Con makes no challenges. Pro wins if Pro satisfies (∗1).

(b) The case that Con makes finitely many challenges. Con wins if Con satisfies the rule (∗2) and his last challenge is successful.

(c) The case that Con makes infinitely many challenges. In this case, Con wins.

Moreover, putting conditions (b) and (c) together, we say that Con wins if and only if Con satisfies (∗2) and

(∗3) $\forall n$(there exists a last challenge at stage n → the challenge is successful).

This condition can be written as a Π_2^0-formula. Thus, the winning condition of player I in game G is defined as follows: player I wins iff

(player I becomes Pro ∧ (Pro satisfies (∗1) ∨ Con does not satisfies (∗2) ∨ Con does not satisfy (∗3))) ∨
(player I becomes Con ∧ (Pro does not satisfy (∗1)) ∨ (Con satisfies (∗2) ∧ Con satisfies (∗3))).

This condition is $\mathsf{Sep}(\Delta_1^0, \Sigma_2^0)$, since "player I becomes Pro" if player II rejects I's choice y^* at the first stage, which is obviously a Δ_1^0 condition. Now, by $\mathsf{Sep}(\Delta_1^0, \Sigma_2^0)$-$\mathsf{Det}$, game G is determined. As we observed above, player I has no winning strategy, and so player II has a winning strategy τ. Let V be the set of pairs (x, y) such that τ calls for II as Pro to put (x, y) into V at every meaning position which could get challenge otherwise.

To see that V satisfies the axiom of inductive definition, we introduce the notion of simple y-play, following Tanaka.[14] Assume $(z, y) \in V$. We consider a τ-consistent play in which I plays $y^* = y$, τ tells II to accept y^*, and no challenge occurs to the last. We call such a play a *simple y-play*.

Let p be a simple y-play. Let W^p be the pre-well-ordering constructed by Pro in p, i.e.,

$$W^p = \{(w, x) : v((w, x)) = 1 \text{ occurs in } p\}.$$

Clearly, $V \subset W^p$. To show V has the same initial segment (below y) as W^p, we assume for the contrary that there is a $w <_{W^p} y$ (also $<_V y$) and $x <_V y$ such that $W^p_{<w} = V_{<w}$ and $(w, x) \in W^p - V$. Since $(x, w) \notin V$, II (as Pro) asserts $x \not\leq w$ at some meaning position. Then I can challenge to the assertion, and win the game by constructing witness copied from the simple y-play for W^p, which contradicts with the assumption that II has a winning strategy. Hence, V and W^p have the same initial segment below y.

Since W^p satisfies $(*1)$, V is also a pre-well-ordering on its field F and $\forall x \in F$ $(W_x = \Gamma(W_{<x}) \cup W_{<x})$. Finally, we need to show $\Gamma(F) \subset F$. By way of contradiction, we assume that there were $y \in \Gamma(F) - F$. Then consider a play in which I plays $y^* = y$ at the beginning. If II always accepts y, y must belong to F. When II rejects y, I (as Pro) can construct V and a witness for $y \in \Gamma(F)$, so that he wins the game. This is a contradiction. Thus, V satisfies the axiom of Σ^1_1-inductive definition. $\qquad\square$

5. Transfinite Recursion of Inductive Definitions

We introduce an axiom $\mathcal{C}$-IDTR which asserts the existence of sets constructed by transfinite recursion of the inductive definitions (following MedSalem and Tanaka[5]).

Definition 5.1. The formal system $\mathcal{C}$-IDTR$_0$ consists of ACA$_0$ and the following axiom scheme ($\mathcal{C}$-IDTR): for any well-ordering $\preceq$ and $\mathcal{C}$-operator Γ, there exists a transfinite sequence $\langle V^r : r \in \text{field}(\preceq) \rangle$ satisfying the following conditions: for each $r \in \text{field}(\preceq)$,

(1) V^r is a pre-well-ordering on its field $F^r = \text{field}(V^r)$.
(2) $\forall x \in F^r (V^r_x = \Gamma^{F^{\prec r}}(V^r_{<x}) \cup V^r_{<x})$.
(3) $\Gamma^{F^{\prec r}}(F^r) \subset F^r$.

where $V^r_x = \{y \in F^r : y \leq_{V^r} x\}$, $V^r_{<x} = \{y \in F^r : y <_{V^r} x\}$, $F^{\prec r} = \bigcup \{F^{r'} : r' \prec r\}$.

Intuitively, inductive definitions with $\mathcal{C}$-operator Γ are iterated transfinitely along $\preceq$ in the following way. First apply inductive operator $\Gamma^\emptyset$ with the empty parameter to obtain a fixed point F^{r_0}, where r_0 is the $\preceq$-least element. Then, apply $\Gamma^{F^{r_0}_0}$ with parameter

F^{r_0} to obtain a fixed point F^{r_1} with the second $\preceq$-least r_1. Then, apply $\Gamma^{F_0^r \cup F^{r_1}}$ to obtain F^{r_2}. We iterate this procedure transfinitely along well-ordering $\preceq$, and then we obtain the sequence of pre-well-orderings $\langle V^r : r \in \text{field}(\preceq) \rangle$. See the figure.

$$F^r \supset \Gamma^{F^{\prec r}}(F^r)$$
(fixed point)

$$\Gamma^{F^{\prec r}}(\emptyset) \cup \Gamma^{F^{\prec r}}(\Gamma(\emptyset)) \dots$$

$$\Uparrow$$

$$F^{r_1} \supset \Gamma^{F^{r_0}}(F^{r_1})$$
(fixed point)

$$\Gamma^{F^{r_0}}(\emptyset) \cup \Gamma^{F^{r_0}}(\Gamma(\emptyset)) \dots$$

$$\Uparrow$$

$$F^{r_0} \supset \Gamma^{\emptyset}(F^{r_0})$$
(fixed point)

$$\Gamma^{\emptyset}(\emptyset) \cup \Gamma^{\emptyset}(\Gamma^{\emptyset}(\emptyset)) \dots$$

Fig. 3. Pre-well-orderings $\langle V^r : r \in \text{field}(\preceq) \rangle$ constructed with Γ by $\mathcal{C}$-IDTR$_0$

In our previous paper,[4] it is only stated without proof that Σ_1^1-IDTR$_0$ and $\Delta((\Sigma_2^0)_2)$-determinacy are equivalent. Before giving a proof, we here remark that the parameter $F^{\prec r}$ of operator $\Gamma^{F^{\prec r}}$ can be generalized to a set $G^{\prec r}$ defined with $F^{\prec r}$ by arithmetical transfinite recursion, i.e., $G^r = \Gamma_1(F^r) \cup G^{\prec r}$ with arithmetical Γ_1. This is because even if the description of Γ_1 is inserted into the description of Γ, the operator is still Σ_1^1. To be more precise, we can modify Γ to produce a pair (F^r, G^r). This kind of adjustment is often needed to use Σ_1^1-IDTR.

Theorem 5.1. *Over* RCA$_0$, *the following are equivalent.*

(1) $\Delta((\Sigma_2^0)_2)$-Det.

(2) Sep(Δ_2^0, Σ_2^0)-Det.

(3) Σ_1^1-IDTR.

Proof. (1)$\Leftrightarrow$(2) are obvious from Theorem 3.1. We first prove

$(3) \Rightarrow (2)$, and then $(1) \Rightarrow (3)$.

Let $\varphi(f)$ be a $\mathsf{Sep}(\Delta_2^0, \Sigma_2^0)$-game. Thus, there exist a Δ_2^0-formula ψ, a Π_2^0-formula η_0 and a Σ_2^0-formula η such that

$$\forall f(\varphi(f) \leftrightarrow (\psi(f) \wedge (\eta_0(f))) \vee (\neg\psi(f) \wedge (\eta_1(f)))).$$

Since $\psi(f)$ is a Δ_2^0-formula, there are Δ_1^0-formulas $\theta_0(y), \theta_1(y)$ such that

$$\forall f((\psi(f) \leftrightarrow \forall n \exists m > n \theta_0(f[m])) \wedge (\neg\psi(f) \leftrightarrow \forall n \exists m > n \theta_1(f[m]))).$$

Without loss of generality, we may assume $\neg\exists s(\theta_0(s) \wedge \theta_1(s))$ and $\theta_0(\langle\rangle)$ holds. Note that $\langle\rangle$ is the empty sequence.

Now, we define a recursive tree T as follows:

$$T = \{s \in (\mathbb{N}^{<\mathbb{N}})^{<\mathbb{N}} : s(0) \subsetneq s(1) \subsetneq \cdots \subsetneq s(|s| - 1) \wedge$$

$$\forall k < |s|(k \text{ is even } \rightarrow \theta_0(s(k))$$

$$\forall k < |s|(k \text{ is odd } \rightarrow \theta_1(s(k))\}$$

Clearly, T does not have an infinite path. Let $\preceq$ be the Kleene-Blouwer ordering on T. Since T is a well-founded tree, for any $f \in \mathbb{N}^{\mathbb{N}}$ there exists the $\preceq$-least $x \in T$ such that $\cup x = x(|x| - 1) \subset f$. So, we define a Π_1^0-formula $\xi(x, f)$ as follows:

$$\xi(x, f) \leftrightarrow x \in T \wedge \cup x \subset f \wedge \forall y(y \in T \wedge \cup y \subset f \rightarrow x \preceq y).$$

Then, it is easy to see

$$\psi(f) \leftrightarrow \exists x(|x| \text{ is even } \wedge \xi(x, f)),$$

$$\neg\psi(f) \leftrightarrow \exists x(|x| \text{ is odd } \wedge \xi(x, f)).$$

Now, we define $\eta_0'(x, f, Y)$ and $\eta_1'(x, f, Y)$ with parameter Y as follows:

$$\eta_0'(x, f, Y) \equiv ((\xi(x, f) \wedge (\eta_0(f))) \vee \exists n(f[n] \in Y)),$$

$$\eta_1'(x, f, Y) \equiv ((\xi(x, f) \wedge (\eta_1(f))) \vee \exists n(f[n] \in Y)).$$

Clearly, the formula η_0' is Π_2^0 and η_1' is Σ_2^0. Thus, their determinacy is deduced from Σ_1^1-ID. More precisely, the set of sure winning positions for player I (II) in a $\Sigma_2^0(\Pi_2^0)$ game exists by Σ_1^1-ID.[7]

By using Σ_1^1-IDTR, we inductively define the set of sure winning positions for player I in the game φ as follows: if $|x|$ is odd then

$$\overline{W}_x = \{s \in \mathbb{N}^\mathbb{N} : \cup x \subset s \text{ and II has a win. st. in } \eta_0'(x, f, W_{\prec x}) \text{ starting at } s\},$$

and if $|x|$ is even then

$$W_x = \{s \in \mathbb{N}^\mathbb{N} : \cup x \subset s \text{ and I has a win. st. in } \eta_1'(x, f, W_{\prec x}) \text{ starting at } s\},$$

where $W_{\prec x} = \bigcup\{W_y : y \prec x, y \text{ is even }\} \cup \bigcup\{\overline{W}_y^c : y \prec x, y \text{ is odd }\}$. We may identify W_y as $\overline{W}_y^c$.

We set $W = \bigcup\{W_x : x \in T\}$. Then, we prove the following:

$\langle\rangle \in W \to$ I wins the game $\varphi(f)$, and
$\langle\rangle \notin W \to$ II wins the game $\varphi(f)$.

First, we assume $\langle\rangle \in W$. Then, $\langle\rangle \in W_x$ for some $x \in T$. In fact, x must be $\langle\langle\rangle\rangle$, and so $|x|$ is odd, hence $\langle\rangle \notin \overline{W}_x$. By Σ_2^0-determinacy, player I has a winning strategy $\sigma_{\langle\rangle}^x$ in $\eta_0'(x, f, W_{\prec x})$. If a play f obeying $\sigma_{\langle\rangle}^x$ satisfies $\forall n(f[n] \notin W_{\prec x})$, player I wins the original game φ because $\xi(x, f) \wedge \eta_0(f)$ holds. If $\exists n(f[n] \in W_{\prec x})$, letting $f[n] = s$, we have $s \in W_y$ for some $y \in T$. Then, player I has a winning strategy σ_s^y in $\eta_0'(y, W_{\prec y})$ at s if $|y|$ is odd, and μ_s^y in $\eta_1'(y, f, W_{\prec y})$ at s if $|y|$ is even. Thus, player I switches his strategy from $\sigma_{\langle\rangle}^x$ to σ_s^y if $|y|$ is odd and μ_s^y if $|y|$ is even. Continuing this procedure, the switches of strategies occur only finitely many times, because T is well-order. Therefore, player I can eventually win the game $\varphi(f)$.

Suppose $\langle\rangle \notin W$. Let $x = \langle\langle\rangle\rangle$. Since $|x|$ is odd, $\langle\rangle \in \overline{W}_x$. So player II has a winning strategy $\tau_{\langle\rangle}^x$ in $\eta_0'(x, f, W_{\prec x})$ for $(\neg\xi(x, f) \vee \neg\eta_0(f)) \wedge \forall n(f[n]) \notin W_{\prec x}$. If a play f consistent with $\tau_{\langle\rangle}^x$ satisfies $\forall n \neg\theta_1(f[n])$, then by the definition of $\xi(x, f)$, we have $\xi(x, f)$ and so $\neg\eta_0(f)$, which implies $\neg\varphi(f)$. Then, we suppose there exists n such that $\theta_1(f[n])$. Let $s = f(n)$, and $y = \langle\langle\rangle, s\rangle$. So $y \prec x$, and $s \notin W_y$. Now player II has a winning strategy for $\neg\eta_0'(y, f, W_{\prec y})$ at s if $|y|$ is odd, and for $\neg\eta_1'(y, f, W_{\prec y})$ at s if $|y|$ is even. Repeat this argument along the well-ordering $\prec$. This completes (3)$\Rightarrow$(2).

We next show (1)$\Rightarrow$(3). Σ_1^1-IDTR asserts the existence of pre-well-orderings $\langle V^r : r \in \text{field}(\prec)\rangle$ such that $x \in \Gamma^{F^{\prec r}}(V_{\prec x}^r) \cup V_{\prec x}^r$ for all $r \in \text{field}(\prec)$ and $x \in F^r$, and such that $\Gamma^{F^{\prec r}}(V^r) \subset F^r$. We define a $\Delta((\Sigma_2^0)_2)$-game G in the same manner as in the proof of Theorem

4.1 to show the existence of V^r from a winning strategy for player II.

At the beginning of a game, Player I asks a question "$y^* \in F^{r^*}$?" by choosing a pair of numbers (y^*, r^*). After player II answers "Yes" or "No", a subgame starts. Depending on the answer of player II, one of the players becomes Pro and constructs the $\leq y^*$-initial segment of pre-ordering V^{r^*} with witnesses for his assertions and also a sequence of sets $\langle F^r : r \prec r^* \rangle$. The other player becomes Con to watch Pro's construction of pre-ordering V^{r^*} and sets F^r for all $r \prec r^*$, and tries to point out mistakes on them. Note that Pro does not construct a pre-ordering V^r on F^r, unless a challenge is made against F^r. If no challenge occurs, it seems to be almost the same as the $\mathsf{Sep}(\Delta_1^0, \Sigma_2^0)$-game of Theorem 4.1. A major difference is following. If Con challenges to Pro's assertion $x \notin F^r$, then the roles of Pro and Con are switched, and *new Pro* starts a subsubgame to construct the $\leq x$-initial segment of V^r, and *new Con* watches the opponent's construction. Since this kind of challenges occur downwards along the well-ordering $\prec$, the switches of the players occur only finite many times. By the same argument in the proof of Theorem 4.1, Player I cannot have a winning strategy. Thus, by the determinacy of game, player II has a winning strategy. With the winning strategy, the desired sets can be constructed.

Now, to describe details of the Pro/Con subgame, we fix a Σ_1^{1Y}-operator Γ^Y. By the normal form theorem, there exists a recursive relation R such that the following holds:

$$x \in \Gamma^Y(X) \leftrightarrow \exists f \forall z R(f[z], x, X[z], Y[z]).$$

Then, Pro constructs pre-ordering V^{r^*} with witness $\{f_x : x \in \text{field}(V^{r^*})\}$ and sequence of sets $\langle F^r : r \prec r^* \rangle$. Con makes challenges to Pro's construction if needs arise.

For each stage n, Pro and Con play $v(n), f(n), s(n)$ and $c(n), e(n), u(n), g(n)$, respectively, where $v, u, s \in \{0,1\}^{\mathbb{N}}$, $f, g \in \mathbb{N}^{\mathbb{N}}$, and $c, e \in \{\mathbb{N} \cup \{-1\}\}^{\mathbb{N}}$.

Pro	Con
$v(0), f(0), s(0)$	
	$c(0), e(0), u(0), g(0)$
$v(1), f(1), s(1)$	
	$c(1), e(1), u(1), g(1)$
$\vdots$	
	$\vdots$

Except the new functions s and e, functions v, f, c, u and g play the same roles as in Theorem 4.1. Namely, v is used to constructs the pre-ordering $V^{r^*} = \{(n_0, n_1) : v(n) = 1\}$, and f_n is a witness function for each assertion $n_0 \in \Gamma^{F^{\prec r^*}}(V^{r^*}_{<n_1}) \cup V^{r^*}_{<n_1}$ (see (b) below), and u, g give witnesses for Con's assertions. The new function s is used to construct a set $F^{l_1} = \{l_0 : s(l) = 1\}$ for any $l_1 \prec r^*$. Notice that Pro constructed a set $F^r (r \prec r^*)$ rather than a pre-ordering $V^r (r \prec r^*)$. However, if Con makes a challenge on F^r by using e, a new subgame starts with construction of $V^{r'}$ and $F^{r'}(r' \prec r)$. For simplicity of description, the initial stage about "$y^* \in F^{r^*}$?" is often regarded as a challenge by e. Then, after the last e-challenge, Pro is required to satisfy the following conditions, (denote $(*1)$).

(a) V^r is a pre-ordering with y in the field.
(b) $\forall n(v(n) = 1 \to \forall z R(f^n[z], n_0, V^r_{<n_1}[z], F^{\prec r}[z]) \vee n_0 \in V^r_{<n_1})$.

Condition $(*1)$ can be written as a Π^0_1-formula.

In the same way as in Theorem 4.1, Con makes a challenge to a negative assertions $v(n) = 0$ of Pro by playing $c(m) = n$ at stage $m \geq n$. We call this a c-*challenge*. C-challenge must be done along the order already constructed by Pro in the decreasing way, which is called rule $(*2)$ as before. Then, condition $(*3)$ is properly modified as follows (also denote $(*3)$):

$\forall n$(there exists a last c-challenge at stage $n \to$ the c-challenge is successful),

where "Con's c-challenge is successful" if Con constructs an initial segment $U^m = \{k : u(k + m) = 1\} \subset V^r_{<n_1}$ and a witness $g^m(k) = g(k + m)$ such that

$$\forall z R(g^m[z], n_0, U^m[z], F^{\prec r}[z]).$$

In addition, Con makes a challenge to Pro's assertions $s(l) = i \in \{0, 1\}$, which is called an *e-challenge*. If an *e*-challenge is made to $s(l) = 1$ (i.e., Pro asserts $l_0 \in F^{l_1}$), they start a new Pro/Con subgame on F^{l_1} keeping the same roles. If an *e*-challenge is made to $s(l) = 0$, then a new subgame starts where the players change their roles from the initial subgame for constructing the $\leq l_0$-segment of pre-ordering V^{l_1}. *E*-challenges must be made along $\prec$ in the decreasing way, and so may not occur infinite many times.

Comment. The *rules* of challenges get more complicated, but the basic idea is still same as in the proof of theorem 4.1. Challenges must be made in a *decreasing* way. In particular, once an *e*-challenge has made on V^r, then Con is not allowed to make an *e*-challenge to $\langle F^{r'} : r \preceq r' \preceq r^* \rangle$ or a *c*-challenge to $\langle V^{r'} : r \prec r' \preceq r^* \rangle$. As for *c*-challenges, Con must obay the same rule as before. These rules for Con are denoted as (∗∗). If Con violates these rules, then Con looses.

For the other hand, assume that Con has challenged to $v(n_0, n_1) = 0$ for some $r \preceq r^*$. Then, Con can make *c*-challenge to $v(m_0, m_1) = 0$ only if $v(m_1, n_1) = 1$ and $v(n_1, m_1) = 0$ are asserted by Pro before stage m.

Finally, we define the winning conditions of the players. Unlike a $\mathsf{Sep}(\Delta^0_1, \Sigma^0_2)$-game, it is not conclusive for a player to be either Pro or Con at the end. In the following, Pro (Con) means Pro (Con) after the last *e*-challenge.

1. The case that Con makes no challenges. Pro wins iff Pro satisfies (∗1).
2. The case that Con makes finitely many challenges and the last one is an *e*-challenge to $s(l) = \{0, 1\}$. Then, Pro wins iff Pro satisfies (∗1).
3. The case that Con makes finitely many challenges and the last one is a *c*-challenge. Con wins iff Con's challenge is successful.
4. The case that Con makes infinitely many *c*-challenges. In this case, Con wins.

Comment. After *e*-challenge, for Con to win, as satisfying (∗∗), Con makes infinitely many *c*-challenges or, otherwise, there exists the last

c-challenge and Con's challenge is successful. That is,

$(* * *)$ $\forall n($ there exists a last c-challenge $\to$ Con's challenge is successful.$)$

This is clearly a Π_2^0-formula.

So, the winning condition of player I is outlined as follows:

(player I is Pro after the last e-challenge $\to ((*1) \wedge \{\neg(*2) \vee \neg(*3)\}$ holds))

$\wedge$(player I is Con after the last e-challenge $\to (\neg(*1) \vee \{(*2) \wedge (*3)\}$ holds)).

Note that $(*1)$ and $(*2)$ are Π_1^0, but $(*3)$ is Π_2^0. We also remark that e-challenge never occur infinitely many times. So, by ignoring the Π_1^0 conditions, we can rewrite the above condition as follow:

$\exists n($there exists a last e-challenge at stage $n \wedge$ (I is Pro since then $\to \neg(*3)))$

$\wedge \forall n($there exists a last e-challenge at stage $n \to$ (I is Con since then $\to (*3)))$.

Thus, the winning condition of player I is $\Sigma_2^0 \wedge \Pi_2^0$. Similarly, that of player II can be written in $\Sigma_2^0 \wedge \Pi_2^0$. Therefore, this game is a $\Delta((\Sigma_2^0)_2)$-game.

Now, by $\Delta((\Sigma_2^0)_2)$-**Det**, the above game is determined. Since player I cannot have a winning strategy by the same reason as before, player II has a winning strategy τ. For each $r \in \text{field}(\preceq)$, V^r be the set of pair (x, y) such that τ calls for II as Pro to put (x, y) into V^r at every meaningful position which could get challenged otherwise. Then as in the proof of Theorem 4.1, it is easy to see that $\langle V^r \rangle$ satisfies Σ_1^1-**IDTR**.

$\square$

6. Multiple Inductive Definitions

MedSalem and Tanaka[6] introduced inductive definitions with multiple operators to characterize the finite boolean combinations of Σ_2^0 determinacy. Before giving a formed definition, we recall intuitive treatments of multiple inductive definitions.

Suppose Γ_0 has a distinct parameter X_0, and thus also denoted $\Gamma_0^{X_0}$. We first set $X_0 = \emptyset$. Then, repeatedly apply $\Gamma_0^{X_0}$ from the empty set until we get a fixed point, say F_0. Then apply Γ_1 once to get a new $X_0 := \Gamma_1(F_0) \cup X_0$. For the second stage, re-start iterating $\Gamma_0^{X_0}$, and get a fixed point F_1. Then apply Γ_1 once, and so on. The procedure stops when we get a fixed point F such that if F' is the least fixed point of Γ_0^F then $\Gamma_1(F') \subset F$. This can be depicted as follows:

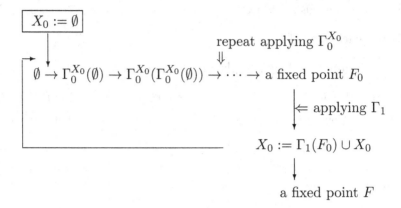

In the above chart, we may start with $X_0 = $ any set Y. So, by $[\Gamma_0, \Gamma_1]^Y$, we denote the fixed point (or its construction) of multiple inductive definitions starting with Y. Now, in general, the combination $[\Gamma_0, \cdots, \Gamma_{k-1}, \Gamma_k]$ can be described as follows. We iterate $[\Gamma_0, \cdots, \Gamma_{k-1}]^\emptyset$ until we get a fixed point, say F_0. Then restart iterating $[\Gamma_0, \cdots, \Gamma_{k-1}]^{\Gamma_k(F_0)}$, get another fixed point F_1, again iterate $[\Gamma_0, \cdots, \Gamma_{k-1}]^{\Gamma_k(F_0) \cup \Gamma_k(F_1)}$ and so on. Eventually, we stop when we get a fixed point F such that if F' is the least fixed point of $[\Gamma_0, \cdots, \Gamma_{k-1}]^F$ then $\Gamma_k(F') \subset F$.

Richter and Aczel[10] already introduced a similar notion of multiple inductive definitions. In fact, translations between the two notions are possible and given in Tanaka[15] (after Definition 2.1 and in the proof of Lemma 3.2.) However, in order to characterize the determinacy of Boolean combinations of Σ_2^0-formulas, the present definition seems to suit well.

We now give a formal definition for the axiom of inductive definitions with two operators, denote $[\mathcal{C}_0, \mathcal{C}_1]$-ID.

Definition 6.1. Let $\mathcal{C}_0$ and $\mathcal{C}_1$ are collections of operators. The axiom scheme $[\mathcal{C}_0, \mathcal{C}_1]$-ID$_0$ asserts the following. For any $\Gamma_0 \in \mathcal{C}_0, \Gamma_1 \in \mathcal{C}_1$, there exist $W \subseteq F_1 \times F_1$, and $\langle V^x : x \in F_1 \cup \{\infty\} \rangle$ such that the following are all satisfied.

(1) W is a pre-well-ordering on F_1.

(2) $\forall x \in F_1 \cup \{\infty\}$

- V^x is a pre-well-ordering on its field F_0^x.
- $V_y^x = \Gamma_0^{W_{<x}}(V_{<y}^x) \cup V_{<y}^x$ for all $y \in F_0^x$.
- $W_x = \Gamma_1(F_0^x) \cup W_{<x}$.
- $\Gamma_0^{W_{<x}}(F_0^x) \subset F_0^x$.

(3) $W_\infty = W_{<\infty} = F_1$.

Here, $W_x = \{y : (x,y) \in W\}$, $W_{<x} = \{y : (y,x) \in W$ and $(x,y) \notin W\}$, $V_y^x = \{z : (z,y) \in V^x\}$, and $V_{<y}^x = \{z : (z,y) \in V^x, (y,z) \notin V^x\}$. We also write $[\mathcal{C}]^2$-ID for $[\mathcal{C},\mathcal{C}]$-ID. Similarly, we can define $[\mathcal{C}]^{n+1}$-ID as $[[\mathcal{C}]^n, \mathcal{C}]$-ID.

The following figure shows pre-well-orderings construction by Γ_0 and Γ_1.

$$\cdots \qquad \Gamma_1$$
$$\Gamma_0^{F_1}(\emptyset) \cup \Gamma_0^{F_1}(\Gamma_0^{F_1}(\emptyset)) \cup \cdots = F_0^\infty \qquad F_1 \supset \Gamma_1(F_0^\infty)$$

$$\Uparrow$$

$$\cdots \qquad V^x \quad \Gamma_1 \qquad x$$
$$\Gamma_0^{W_{<x}}(\emptyset) \cup \Gamma_0^{W_{<x}}(\Gamma_0^{W_{<x}}(\emptyset)) \cup \cdots = F_0^x \qquad W_{<x}$$

$$\Uparrow$$

$$\cdots \qquad V^{x_1} \quad \Gamma_1 \quad x_1$$
$$\Gamma_0^{W_{x_0}}(\emptyset) \cup \Gamma_0^{W_{x_0}}(\Gamma_0^{W_{x_0}}(\emptyset)) \cup \cdots = F_0^{x_1} \qquad W_{x_0} = W_{<x_0} = \Gamma_1(F_0^{x_0})$$
$$\Uparrow$$

$$\cdots \qquad V^{x_0} \quad \Gamma_1 \qquad x_0$$
$$\Gamma_0^{\emptyset}(\emptyset) \cup \Gamma_0^{\emptyset}(\Gamma_0^{\emptyset}(\emptyset)) \cup \cdots = F_0^{x_0}$$

Fig. 4. Pre-well-orderings constructed by Γ_0 and Γ_1

MedSalem and Tanaka[6] showed the equivalence of $[\Sigma_1^1]^2$-ID to $(\Sigma_2^0)_2$-Det. Considering $\mathsf{Sep}(\Delta_1^0, (\Sigma_2^0)_2)$-Det, which turns out to be equivalent to $[\Sigma_1^1]^2$-ID, we can give simple but more rigorous arguments for their equivalence.

Theorem 6.1 (cf. MedSalem-Tanaka[6]). *Over* ACA_0, *the following are equivalent.*

(1) $(\Sigma_2^0)_2$-Det.
(2) $\mathsf{Sep}(\Delta_1^0, (\Sigma_2^0)_2)$-Det.
(3) $[\Sigma_1^1]^2$-ID.

Proof. $(1)\Rightarrow(2)$. View a $\mathsf{Sep}(\Delta_1^0, (\Sigma_2^0)_2)$-game as a Δ_1^0-game with an oracle telling which player has a winning strategy for the $(\Sigma_2^0)_2$ (or $(\Pi_2^0)_2$)-game starting at each position. Note that we may use Δ_2^1-CA_0 to define the oracle, since it is Π_3^1-conservative over Π_1^1-CA_0 (see the proof of Theorem 4.1).

$(3)\Rightarrow(1)$. Let $A(f)$ be a $(\Sigma_2^0)_2$-formula of the form $\exists x \forall y R(x, f[y]) \wedge \psi(f)$, where ψ is Π_2^0 and R is Π_0^0. Intuitively, we define a transfinite sequence $\langle W_\beta : \beta < \alpha \rangle$ of sure winning positions for player I as follows: for any ordinal $\beta < \alpha$,

$$u \in W_\beta \leftrightarrow \underbrace{\exists x}_{(1)} \underbrace{(\text{player I has a winning strategy for } A_{u,\beta,x})}_{(2)}$$

where $A_{u,\beta,x}(f) \equiv \forall y (R(x, (u * f)[y]) \vee (u * f)[y] \in W_{<\beta}) \wedge \psi(u * f)$, $u * f$ denotes the concatenation of u and f, and $W_{<\beta} = \bigcup_{\gamma < \beta} W_\gamma$.

Here, part (1) of the right hand side of this definition is a Σ_1^0-operator. Part (2) can be regarded as the complement of the fixed point of Σ_1^1-operator. Thus, $\langle W_\beta \rangle$ is defined by a combination of a Σ_1^0 operator and a Σ_1^1-operator. Hence, $W_\alpha = \bigcup_{\beta < \alpha} W_\beta$ exists by $[\Sigma_1^1]^2$-ID.

Then it is easy to see that

(1) $u \in W_\alpha \to$ player I has a winning strategy for A^u,
(2) $u \notin W_\alpha \to$ player II has a winning strategy for A^u,

where A^u is defined by $f \in A^u \leftrightarrow u * f \in A$. See MedSalem-Tanaka[6] for the details.

$(2)\Rightarrow(3)$. $[\Sigma_1^1]^2$-ID asserts the existence of a pre-well-ordering W and $\langle V^x : x$ in the field of $W \rangle$. A major difference between Σ_1^1-IDTR and $[\Sigma_1^1]^2$-ID is that a well-ordering $\preceq$ is given on an assumption in Σ_1^1-IDTR, while the corresponding pre-well-ordering W must be constructed by an inductive operator in $[\Sigma_1^1]^2$-ID. In what follows, taking over the notation and terms for game constructions from the previous theorems, we put emphasis on additional machinery.

We now define a $\mathsf{Sep}(\Delta_1^0, (\Sigma_2^0)_2)$-game to handle $[\Sigma_1^1]^2$-ID. As in the proofs of the previous theorems, player I starts the game by asking a question "$y^* \in F_1$?" Then, player II answers, and a Pro/Con subgame starts, where Pro constructs a pre-ordering W with y^* in its field and its witness f, a pre-well-ordering V^{y^*}, and a sequence of sets F_0^x for any $x \in F_1$. At each stage n, Pro and Con play $v(n), w(n), f(n), s(n)$ and $c(n), d(n), u(n), t(n), g(n), e(n)$, respectively. Functions v, f, s, c, u, g, and e are used in the same way as in the proof of Theorem 5.1. As for the new functions, w constructs pre-ordering $W_{<y^*}$, d is used to make a challenge to $w(m) = 0$ (called d-challenge), and t constructs $T \subseteq W_{<m_1}$ after a d-challenge. We need treat the winning conditions concerning d-challenges. Let $d(n) = m$. This means that Con makes challenges to $m_0 \notin \Gamma_1(W_{<m_1})$. We suppose

$$x \in \Gamma_1(X) \leftrightarrow \exists f \forall z R_1(f[z], x, X[z])$$

where R_1 is recursive. Then, "Con's d-challenge is successful" if he constructs an initial segment $T^n \subseteq W_{<m_1}$ and a witness g^n such that

$$\forall z R_1(g^n[z], m_0, T^n[z]).$$

This can be written as a Π_1^0-formula. Note that there may be no l such that $T = W_{<l}$, since $W_{<m_1}$ may not be well-founded. Then Con(at the end)wins if Con satisfies $(*2)$ and $(*3'')$the last d-challenge is successful (if any).

Also, we need a small modification on the definition of "Con's c-challenge is successful." Since Con's c-challenge $c(n) = m$ is made to $m_0 \notin \Gamma_0^{W_{<m_2}}(V_{<m_1}^{m_2})$, "Con's c-challenge is successful" if he constructs an initial segment $U^n \subset V_{<m_1}^{m_2}$ and witness g^n such that

$$\forall z R_0(g^n[z], m_0, U[z], W_{<m_2}[z]),$$

where $m = (m_0, m_1, m_3)$.

An important difference between the game of Theorem 5.1 and one under construction is the treatment of e-challenges. In the previous game, e-challenges are made along a given well-order $\prec$, and so may not occur infinitely many times. However, in this game, e-challenge are made along W, which may not be well-ordered. But, if e-challenges occur infinitely many times, Pro at the beginning loses the game by rule.

To outline the winning conditions, we again ignore the Π_1^0 conditions. Then, we consider the following four conditions:

(p_1) $\exists n$ (there exists a last e-challenge at stage n $\wedge$ (player I is Pro $\rightarrow \neg(*3)$)).

(p_2) $\forall n$ (there exists a last e-challenge at stage n $\rightarrow$ (player I is Con $\rightarrow (*3)$)).

(c_1) $\exists n$ (there exists a last e-challenge at stage n $\wedge$ player I is Pro $\wedge\neg(*3)$).

(c_2) $\forall n$ (there exists a last e-challenge at stage n $\rightarrow$ (player I is Con $\wedge(*3)$)).

Then, "player I wins" iff

(player I starts a subgame as Pro $\wedge$ $((p_1) \wedge (p_2))$) $\vee$
(player I starts a subgame as Con $\wedge$ $((c_1) \vee (c_2))$).

Clearly, the above condition is $\mathsf{Sep}(\Delta_1^0, (\Sigma_2^0)_2)$. We also remark that if e-challenge occur infinitely many time, then (c_2) holds. So by $\mathsf{Sep}(\Delta_1^0, (\Sigma_2^0)_2)$-$\mathsf{Det}$, player II has a winning strategy τ. Let W and $\langle V^x \rangle$ be the structures where τ calls for II as Pro to construct. Then as in the previous proofs, it is easy to see that they satisfy $[\Sigma_1^1]^2$-ID.

$\square$

We now introduce a new axiom scheme $[\Sigma_1^1]^2$-IDTR_0 by combining transfinite recursion with multiple inductive definitions, and prove that it is equivalent to $\Delta((\Sigma_2^0)_3)$-Det.

Definition 6.2. Let $\mathcal{C}_0$ and $\mathcal{C}_1$ are collections of operators. The axiom scheme $[\mathcal{C}_0, \mathcal{C}_1]$-$\mathsf{IDTR}_0$ asserts the following. For any well-ordering $\preceq$ and any $\Gamma_0 \in \mathcal{C}_0, \Gamma_1 \in \mathcal{C}_1$, there exist $\langle W^r : r \in \text{field}(\preceq) \rangle$, $\langle V^{r,x} : r \in \text{field}(\preceq), x \in F_1^r \rangle$ and $\langle V^{r,\infty} : r \in \text{field}(\preceq) \rangle$ such that the following are all satisfied.

(1) W^r is a pre-well-ordering on its field F_1^r.
(2) $\forall x \in F_1^r \cup \{\infty\}$

- $V^{r,x}$ is a pre-well-ordering on its field $F_0^{r,x}$.

- $V_y^{r,x} = \Gamma_0^{F_1^{\prec r} \oplus W_{<x}^r}(V_{<y}^{r,x}) \cup V_{<y}^{r,x}$ for all y $\in F_0^{r,x}$.
- $W_x^r = \Gamma_1^{F_1^{\prec r}}(F_0^{r,x}) \cup W_{<x}^r$.
- $\Gamma_0^{F_1^{\prec r} \oplus W_{<x}^r}(F_0^{r,x}) \subset F_0^{r,x}$.

(3) $W_\infty^r = W_{<\infty}^r = F_1^r$.

where $F_1^{\prec r} = \oplus\{F_1^{r_i} : r_i \prec r\} = \{(r_i, x) : x \in F_1^{r_i}, r_i \prec r\}$. Note also that $X \oplus Y = \{2x : x \in X\} \cup \{2y + 1 : y \in Y\}$.

We see how the axiom works intuitively. We first assume that $W^{\prec r}$ has been constructed and then consider the rth construction of $[\mathcal{C}]^2$-IDTR$_0$. $\Gamma_0^{F_1^{\prec r}}$-operator is applied until we get the fixed point F_0^{r,x_0}. Then, $\Gamma_1^{F_1^{\prec r}}(F_0^{r,x_0}) = W_{x_0}^r$ is constructed. Next, $W_{x_0}^r$ is joined to $F_1^{\prec r}$, and we obtain $\Gamma_0^{F_1^{\prec r} \oplus W_{x_0}^r}$. This procedure is repeated until F_1^r becomes the common fixed point of those two operators. Note that the pre-well-ordering constructed by $[\mathcal{C}]^2$-ID$_0$ is equal to the r_0th (i.e. the first) construction by $[\mathcal{C}]^2$-IDTR$_0$, where r_0 is the $\preceq$-least element of field($\preceq$).

Then, we have the following theorem.

Theorem 6.2. *Over* ACA$_0$, *the following are equivalent.*

(1) $\Delta((\Sigma_2^0)_3)$-Det.
(2) Sep($\Delta_2^0, (\Sigma_2^0)_2$)-Det.
(3) $[\Sigma_1^1]^2$-IDTR$_0$.

Proof. (1)$\Leftrightarrow$(2) is obvious from Theorem 3.1. (3)$\Rightarrow$(2) can be proved in the same way as in the proof of (3)$\Rightarrow$(2) of Theorem 5.1.

We show (1)$\Rightarrow$(3). $[\Sigma_1^1]^2$-IDTR$_0$ asserts the existence of sequences of pre-well-orderings $\langle W^r : r \in \text{field}(\preceq)\rangle$ and $\langle V^{r,x} : r \in \text{field}(\preceq), x \in W^r\rangle$. We define a $\Delta((\Sigma_2^0)_3)$-game in the same manner as before, and we will highlight its difference from the Sep($\Delta_1^0, (\Sigma_2^0)_2$)-game.

To check the TR part, Con needs one more challenge function, say e_1, than the proof of $[\Sigma_1^1]^2$-ID$_0$. As in the proofs of Theorem 5.1 and 6.1, after (y^*, r^*) are chosen, Pro constructs pre-ordering $W_{y^*}^{r^*}$ and V^{r^*,y^*} with functions w and v, respectively. Pro also constructs a sequence of sets $\langle F_0^{r,x} : r \preceq r^*, x \in F_1^r\rangle$ by the function s. To do so, we identify a triple (n_0, n_1, n_2) as its code n such that $F_0^{n_1,n_2} =$

$\{n_0 : s(n) = 1\}$. In addition to the function s, we prepare another function $s_1 : \mathbb{N} \to \{0,1\}$ to construct a sequence $\langle F_1^r : r \prec r^* \rangle$ such that $F_1^{n_2} = \{n_0 : s_1(n) = 1\}$. Then, the function e_1 is used to make a challenge by playing $e_1(m) = n$ to a Pro's assertion $s_1(n) = i \in \{0,1\}$ for $n \leq m$. Since the TR is iterated along the given well-ordering, we know there are no infinitely many e_1-challenges. Thus, there must exist the last e_1-challenge. After the last e_1-challenge, the $\Delta((\Sigma_2^0)_3)$-game can be reduced to a $\mathsf{Sep}(\Delta_1^0, (\Sigma_2^0)_2)$-game. This condition for player I is $(\Pi_2^0)_3$, and so is for player II. Therefore, this game is $\Delta((\Sigma_2^0)_3)$.

Now, to describe the winning conditions, we consider the following conditions.

(c_1') $\forall n$ (there exists a last e_1-challenge at stage n $\to$ player I is Con $\wedge$ (c_2)).

(c_2') $\forall n$ (there exists a last e_1-challenge at stage n $\to$ (player I is Pro $\to$ (p_2)))).

(c_3') $\exists n$ (there exists a last e_1-challenge at stage n $\wedge$ (player I is Pro $\to$ (p_1)) $\wedge$ (player I is Con $\to$ (c_1)))).

Then, player I wins if a $(\Pi_2^0)_3$ condition $(c_1') \vee \{(c_2') \wedge (c_3')\}$ holds. Similarly, the winning condition of player II can be written in $(\Pi_2^0)_3$. Therefore the game is $\Delta((\Sigma_2^0)_3)$, and as determinate. Thus, the structures we wants are obtained from II's winning strategy. $\square$

Theorem 6.3. *Over* ACA_0, *the following are equivalent.*

(1) $(\Sigma_2^0)_3$-Det.
(2) $\mathsf{Sep}(\Delta_1^0, (\Sigma_2^0)_3)$-$\mathsf{Det}$.
(3) $[\Sigma_1^1]^3$-ID_0.

Proof. $(3) \Rightarrow (1) \Leftrightarrow (2)$ can be proved as before. Thus, we only consider $(2) \Rightarrow (3)$. In this game, e_1-challenges can be made infinitely many times because the new operator Γ_2 will construct a pre-ordering. So, after the last e_1-challenge (if any), the game is essentially same as a $\Delta((\Sigma_2^0)_3)$-game.

We constructs a $\mathsf{Sep}(\Delta_1^0, (\Sigma_2^0)_3)$-game such that "player I wins" iff

(player I starts a subgame as Con) $\wedge$ $((c_1') \vee ((c_2') \wedge (c_3')))$ $\vee$
(player I starts a subgame as Pro) $\wedge$ $((p_1') \wedge ((p_2') \vee (p_3')))$.

where (c_1'), (c_2'), (c_3') are already given, (p_1'), (p_2'), (p_3') are the negations of (c_1'), (c_2'), and (c_3'), respectively, with Con and Pro switched. In other words, (p_1'), (p_2') and (p_3') are stated as follows:

(p_1') $\exists n$(a last e_1-challenge exists at n $\wedge$(player I is Con $\rightarrow$ $\exists m > n$(a last e-challenge exists at m $\wedge$ (player I is Pro $\rightarrow$ $\neg(*3)$))))).

(p_2') $\exists n$(a last e_1-challenge exists at stage n $\wedge$ (player I is Pro $\wedge$ $\exists m > n$(a last e-challenge exists $\wedge$(player I is Pro $\wedge \neg(*3)$))))).

(p_3') $\forall n$(a last e_1-challenge exists $\rightarrow$(player I is Pro $\wedge$ $\forall m > n$(a last e-challenge exists $\rightarrow$(player I is Con $\wedge(*3)$)))) $\vee$ (player I is Con $\wedge$ $\forall m > n$(a last e-challenge exists $\rightarrow$ (player I is Con $\rightarrow (*3)$))))).

Compare also this construction with that of the $\mathsf{Sep}(\Delta_1^0, (\Sigma_2^0)_2))$-game. $\square$

We then proceed to $[\Sigma_1^1]^3$-IDTR, and prove that it is equivalent to $\Delta((\Sigma_2^0)_4)$-Det. Since, $[\Sigma_1^1]^k$-IDTR is just $\mathcal{C}$-IDTR with $\mathcal{C} = [\Sigma_1^1]^k$, we omit the formal definition of $[\Sigma_1^1]^k$-IDTR.

Theorem 6.4. *Over* ACA_0, *the following are equivalent.*

(1) $\Delta((\Sigma_2^0)_4)$-Det.
(2) $\mathsf{Sep}(\Delta_2^0, (\Sigma_2^0)_3)$-Det.
(3) $[\Sigma_1^1]^3$-IDTR.

Proof. (1)$\Leftrightarrow$(2)$\Leftarrow$(3) are routine. For (1)$\Rightarrow$(3), we construct a game where Con needs one more challenge function e_2 than the proof of $[\Sigma_1^1]^3$-ID.

The winning condition of player I is defined as follows: Player I wins if and only if

$\forall n(E_2 \rightarrow$ (player I is Con) $\wedge$ $(c_1'))\vee$
$[\exists n(E_2 \wedge$(player I is Con $\rightarrow$ $(c_3'))\wedge$(player I is Pro $\rightarrow (p_1')))$ $\wedge$
$\{\forall n(E_2 \wedge$(player I is Con $\rightarrow$ $(c_2'))\wedge$(player II is Pro $\rightarrow (p_3')))$ $\vee$
$\exists n(E_2 \wedge$(player II is Pro $\wedge(p_2')))\}]$.

where E_2 denotes "there exists a last e_2-challenge at stage n".

Since e_2-challenges are made along a given well-ordering, the e_2-challenges can be made only finitely many times. Thus, after the last e_2-challenge, this game can be reduced to a $\mathsf{Sep}(\Delta_1^0, (\Sigma_2^0)_3)$-game. Compare this with the construction of the $\Delta((\Sigma_2^0)_3)$-game. Thus, the winning condition of player I is $(\Pi_2^0)_4$, and so is player II's. Therefore, this game is $\Delta((\Sigma_2^0)_4)$. The desired structures are obtained from II's winning strategy. □

Finally, by induction, we conclude:

Theorem 6.5. *Over* RCA_0, *the following are equivalent. For any* $k > 0$,

(1) $(\Sigma_2^0)_k$-Det.
(2) $\mathsf{Sep}(\Delta_1^0, (\Sigma_2^0)_k)$-$\mathsf{Det}$.
(3) $[\Sigma_1^1]^k$-ID_0.

Theorem 6.6. *Over* RCA_0, *the following are equivalent. For any* $k > 0$,

(1) $\Delta((\Sigma_2^0)_{k+1})$-$\mathsf{Det}$.
(2) $\mathsf{Sep}(\Delta_2^0, (\Sigma_2^0)_k)$-$\mathsf{Det}$.
(3) $[\Sigma_1^1]^k$-IDTR_0.

7. Conclusion and Future Studies

In this paper, we introduced the axiom of transfinite recursion of Σ_1^1 inductive definitions with k operators, denoted $[\Sigma_1^1]^k$-IDTR_0, and showed that it is equivalent to the determinacy of $\Delta((\Sigma_2^0)_{k+1})$ sets. A key fact used in the proof is that a $\Delta((\Sigma_2^0)_{k+1})$ set is expressed as a $\mathsf{Sep}(\Delta_2^0, (\Sigma_2^0)_k)$ set, namely a Δ_2^0-separated union of a $(\Sigma_2^0)_k$ set and $(\Pi_2^0)_k$ set. By virtue of this fact, we can utilize a difference hierarchy for a Δ_2^0 set, to construct a winning strategy for a $\Delta((\Sigma_2^0)_{k+1})$-game.

MedSalem and Tanaka[6] have pinned down the exact determinacy strength of Δ_3^0 sets in terms of transfinte combinations of Σ_1^1 inductive definitions. We should notice that their axiom for transfinte combinations of Σ_1^1 inductive definitions is much stronger than our $[\Sigma_1^1]^k$-IDTR here. However, it is worth studying such an axiom as $[\Sigma_1^1]^\alpha$-IDTR, where α is an ordinal, to refine their result on Δ_3^0-games.

Bradfield[1] has shown that the sets of Player I's winning positions of a $(\Sigma_2^0)_k$-game are exactly the same as the $(k+1)$-level of μ-calculus alternation hierarchy Σ_{k+1}^μ. Then, Bradfield[2] claims that the hierarchy $\langle \Sigma_n^\mu, n \in \omega \rangle$ is strict, that is, for any k in ω, we have $\Sigma_k^\mu \subsetneqq \Sigma_{k+1}^\mu$. This result easily follows from the previous result on multiple inductive definitions[6] together with observation that for any k in ω, Π_2^1-CA_0 proves the consistency of Δ_2^1-$\mathsf{CA}_0 + (\Sigma_2^0)_k$-$\mathsf{Det}$, while it does not prove the consistency of $(\Sigma_2^0)_{<\omega}$-Det_0.

From the main result of this paper, we will also obtain the following refinement. First of all, the hierarchy $\langle \Pi_n^\mu, n \in \omega \rangle$ is naturally defined and so is $\langle \Delta_n^\mu, n \in \omega \rangle$. Then, by the argument of this paper, we can associate a Δ_{n+1}^μ-formula with transfinite recursion of a Σ_k^μ-formula. Moreover, for any k in ω, we have $\Sigma_k^\mu \subsetneqq \Delta_{k+1}^\mu \subsetneqq \Sigma_{k+1}^\mu$ by a similar observation as above. Details will appear in the future literature.

A very earlier version of this paper was presented in a workshop[4] with K. Mashiko as a co-author. A later version was also reported by the present two authors in the conference "Computability in Europe 2012." The authors would like to appreciate the opportunity to participate in them.

References

1. J.C. Bradfield, Fixpoints, games and the difference hierarchy, *Theor. Inform. Appl.* 37, 1-15 (2003).
2. J.C. Bradfield, The modal μ-calculus alternation hierarchy is strict, *Theor. Comput. Sci.* 195, 133-153 (1998).
3. C. Heinatsch, M. Möllerfeld, The determinacy strength of Π_2^1-comprehension, *Ann. Pure Appl. Logic* **161**, 1462-1470 (2010).
4. K. Mashiko, K. Tanaka, K. Yoshii, Determinacy of the Infinite Games and Inductive Definition in Second Order Arithmetic, *RIMS Kokyuroku*, **1729**, 167-177 (2011).
5. M.O. MedSalem, K. Tanaka, Δ_3^0-determinacy, comprehension and induction, *Journal of Symbolic Logic*, **72**, 452-462 (2007).
6. M.O. MedSalem, K. Tanaka, Weak determinacy and iterations of inductive definitions, in Chitat Chong et al. (ed.) *Computational Prospects of Infinity, Part II: Presented talks*, World Sci. Publ., 333-353 (2008).
7. Y.N. Moschovaskis, Descriptive Set theory, *North Halland* (1980).

8. M. Montalbán, R.A. Shore, The Limits of determinacy in second order arithmetic, *Proc. Lond. Math. Soc.* (3) 104, no. 2, 223-252 (2012).

9. T. Nemoto, M.O. MedSalem, K. Tanaka, Infinite games in the Cantor space and subsystems of second order arithmetic, *Math. Log. Quart.*, 53, 226-236 (2007).

10. W. Richter, P. Aczel, Inductive definitions and reflecting properties of admissible ordinals, J.E. Fenstad and P.G. Hinman, eds., General Recursion Theory, 301-381 (1974).

11. S.G. Simpson, *Subsystems of Second Order Arithmetic*, Springer (1999).

12. J.R. Steel, Determinateness and subsystems of analysis, Ph.D. Thesis, Berkeley, (1977).

13. K. Tanaka, Weak axioms of determinacy and subsystems of analysis I (Δ_2^0 games), *Z. Math. Logik Grundlag. Math.*, 36, 481-491 (1990).

14. K. Tanaka, Weak axioms of determinacy and subsystems of analysis II (Σ_2^0 games), *Ann. Pure Appl. Logic* 52, 181-193 (1991).

15. K. Tanaka, A note on multiple inductive definitions, 10th Asian Logic Conference, 345-352, World Sci. Publ., Hackensack, NJ, (2010).

16. P.D. Welch, Weak Systems of determinacy and arithemtical quasi-inductive definitions, *J. Symbolic Logic* 76, no. 2, 418-436 (2011).

A Survey of the Distributional Complexity for AND-OR Trees

Weiguang Peng

School of Mathematics and Statistics, Southwest University, China
E-mail: pwgedu@swu.edu.cn

Liu and Tanaka (2007) first investigated the eigen-distribution, which achieves the distributional complexity for uniform binary AND-OR trees. Subsequently, several works have been done on optimal algorithms and distributional complexity on different kinds of game trees. These works contribute to the studies on the query complexity of game trees with respect to different kinds of distributions: correlated distribution, independent distribution. Here, we give a short survey on these works.

Keywords: Game Trees; Analysis of Algorithms; Independent Distribution; Computational Complexity.

1. Introduction

The AND-OR (OR-AND) game tree is known as a representative model for Boolean expressions. One fundamental issue on AND-OR trees is the evaluation problem, namely, how to decide the *value of a tree* (the value of the corresponding Boolean function) with the smallest cost. The cost to compute a tree on an assignment ω under an algorithm A, denoted as $C(A, \omega)$, is defined as the total number of the leaves that are queried during the execution of algorithm A. It is known that for AND-OR trees, in the worst case, the deterministic algorithm must read all the leaves to determine the value of a tree.

A randomized algorithm is a distribution over a family of deterministic algorithms. For a randomized algorithm, cost is computed as the expected cost over the corresponding family of deterministic algorithms. Yao's principle[14] indicates the relation between randomized complexity and distributional complexity as follows,

$$\underbrace{\min_{A_R} \max_{\omega} cost(A_R, \omega)}_{\text{Randomized complexity}} = \underbrace{\max_{d} \min_{A_D} cost(A_D, d)}_{\text{Distributional complexity}}.$$

where A_R ranges over randomized algorithms, ω ranges over assignments for leaves, d ranges over distributions on assignments and A_D ranges over deterministic algorithms. This result provides a new perspective to analyze randomized algorithms. Saks and Wigderson[9] showed that for any n-branching tree, the randomized complexity is

$$\Theta\left(\left(\frac{n-1+\sqrt{n^2+14n+1}}{4}\right)^h\right),$$

where h is the height of tree. Tarsi[13] studied the evaluation of AND-OR trees with respect to independent and identical distributions and proved that a depth-first algorithm is optimal for balanced AND-OR trees. By Yao's principle,[14] instead of computing the lower bound on the cost of a randomized algorithm, it is enough to treat deterministic algorithms with respect to a distribution over assignments.

Subsequently, Liu and Tanaka[3] characterized the *eigen-distribution* that achieves the distributional complexity, namely a distribution δ over assignments for leaves which satisfies

$$\min_{A} C(A, \delta) = \underbrace{\max_{d} \min_{A} C(A, d)}_{\text{distributional complexity}},$$

where A runs over all deterministic algorithms, d over distributions on assignments for leaves, and $C(A, d) = \Sigma_\omega d(\omega) C(A, \omega)$. Then they launched the study to calibrate eigen-distributions for different sorts of trees on different kinds of distributions (CD for correlated, ID for independent, or IID for independent and identical) under different classes of algorithms (e.g., directional or non-directional).

In the CD case, Liu and Tanaka[3,4] introduced the notion of i-set and E^i-distribution. Then they proved that for a uniform binary tree, E^i-distribution is the unique eigen-distribution, that is, the uniform distribution on the i-set (the Liu-Tanaka theorem). Suzuki and Nakamura[10] studied variants of Liu-Tanaka theorem for some classes of algorithms in the context of uniform binary trees. Peng et al.[7] extended their results to balanced multi-branching trees, and they showed the equivalence between the eigen-distribution and E^i-distribution for such trees. It had remained open that what is the relationship between eigen-distribution and E^i-distribution for generalized game trees, particularly for weighted trees. The weighted tree is first investigated by Saks and Wigderson in,[9] where the cost weight of evaluating each leaf is no long fixed, but associated with its index for leaves and its Boolean value. Okisaka et al.[5] concentrated on a week version of weighted tree, where the cost only depends on the boolean value, which is called balanced multi-branching trees weighted with (a, b). Here, trees with weight $a > 0$ for the leaves with value 1 and $b > 0$ for value 0. If we take $a = b = 1$, these trees are nothing but usual ones with the unit weight.

In the ID case, Liu and Tanaka[3] claimed that for any uniform binary AND-OR tree, if an ID achieves the equilibrium, it turns out to be an IID. Suzuki and Niida[11] proved the claim for uniform binary trees under depth-first algorithms, first fixing the probability r of the root being 0 in the interval $(0, 1)$ and then releasing this condition. Subsequently, Peng et al.[8] extended it to balanced multi-branching trees with the same probability condition. Note that a tree is called *balanced* if the non-terminal nodes of the same height have the same number of child nodes and all the leaves have the same height, the balancedness makes no restriction on the number of children for nodes at different levels. We denote the n-branching tree with height h by $\mathcal{T}_n^h$. Suzuki[12] showed that Peng et al.'s results still holds while non-depth-first algorithms are considered. In,[15] we showed that one algorithm DIR_d is optimal among all depth-first algorithms for any weighted tree, we generalize the previous results in Suzuki[12] to weighted trees with (a, b).

2. Preliminary

In this study, we restrict ourselves to alpha-beta pruning algorithms. It should be noted that such a algorithm is both depth-first and deterministic. Depth-first means that when the algorithm evaluates the value of a certain node, it would not stop querying the leaves under this node until it knows the value of the node. An algorithm is directional if it queries the leaves in a fixed order, independent from the query history.[6] A typical directional algorithm SOLVE evaluates a tree from left to right.[6] otherwise, it is called a non-directional algorithm. We denote $\mathcal{A}_D$ the set of all alpha-beta pruning algorithms, and $\mathcal{A}_{dir}$ the set of all directional algorithms.

We often identity "node" with "node-code".

In this paper, we mainly treat AND-OR trees, in which root is labelled by AND (OR), internal nodes are alternatively labelled by either OR or AND, and leaves are associated with Boolean values 1 and 0. The value of each AND node (each OR node, respectively) is evaluated as the minimum (the maximum, respectively) among all the values of its children. The value of a tree is the value of its root. The *height* of a node is the length of the path from the root to this node. The height of the root is 0 and the height of a tree is the largest height of the leaves.

An *assignment* for $\mathcal{T}$ is a function ω from the set of leaves to Boolean values $\{0, 1\}$. Note that by identifying each leaf with its value, an assignment ω can also be seen as a 0-1 sequence, whose length is the number of leaves.

Let Ω be the set of assignments for a given tree. We say $d : \Omega \to [0, 1]$ is an *independent distribution* (denoted by $d \in \mathrm{ID}$) if there exist p_i's (the probability of the i-th leaf that has value 0) such that for any $\omega \in \Omega$,

$$d(\omega) = \prod_{\{i:\ \omega(i)=0\}} p_i \prod_{\{i:\ \omega(i)=1\}} (1 - p_i).$$

We say $d \in \mathrm{IID}$ if d is an ID satisfying $p_1 = p_2 = \cdots = p_n$. By segment-wise IID, we mean that all leaves belong to the same subtrees of the root λ have the same probability being 0, there are no restriction on the probability for leaves under different subtrees.

Let $C(A, \omega)$ denote the cost of an algorithm A under an assignment ω. Given a set of assignments Ω, a distribution d on Ω such that $\sum_{\omega \in \Omega} d(\omega) = 1$, and $A \in \mathcal{A}_D$, then the expected cost by A w.r.t. d is defined by $C(A, d) = \sum_{\omega \in \Omega} d(\omega) \cdot C(A, \omega)$. The concept of "transposition" has been introduced to investigate $\mathcal{T}_2^h$ in.[10] We extended this notion to n-branching trees.

First, we define a node-code for $\mathcal{T}_n^h$ as follows.

Definition 2.1 (Node-code). *Given a tree $\mathcal{T}_n^h$, a node-code is a finite sequence over $\{0, 1, \cdots, n-1\}$.*

- *The node-code of root is the empty sequence ε.*
- *For a nonterminal node with node-code v, the node-code for its n children are in the form of $v0, v1, \cdots, v(n-1)$ from left to right.*

Definition 2.2 (Transposition of node, an extension of Definition 4 in[10]). *For $\mathcal{T}_n^h$, suppose u is an internal node. For $i < n$, by $\mathrm{tr}_i^u(v)$, we denote the i-th u-transposition of a node v in $\mathcal{T}_n^h$ (Fig. 1), which is defined as follows*

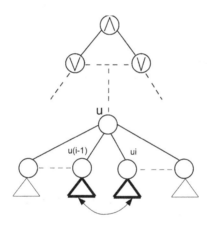

Fig. 1.   Transposition of nodes under node u

- *The 0-th u-transposition of v is itself, that is, $\mathrm{tr}_0^u(v) = v$.*
- *For $i \in \{1, \cdots, n-1\}$, $\mathrm{tr}_i^u(v)$ is defined by*

$$\mathrm{tr}_i^u(v) = \begin{cases} u(i-1)s & \textit{if } v = uis, \\ uis & \textit{if } v = u(i-1)s, \\ v & \textit{otherwise} \end{cases}$$

where s is a finite sequence over $\{0, 1, \cdots, n-1\}$.

Example 2.1. Fig. 2 shows an example of $\mathcal{T}_3^2$ with assignment $\omega = 000100111$. For transposition of node, if $u = 0$ and $i = 1$, then $\mathrm{tr}_1^0(00) = 01$, $\mathrm{tr}_1^0(01) = 00$, and for other v, $\mathrm{tr}_1^0(v) = v$. For transposition of assignment, $\mathrm{tr}_2^\varepsilon(\omega) = 000111100$, and $\mathrm{tr}_1^1(\omega) = 000010111$.

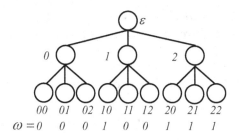

Fig. 2.   An example of $\mathcal{T}_3^2$

Definition 2.3 (Transposition of algorithm). *For $\mathcal{T}_n^h$, suppose that u is an internal node, and $\mathbb{A}$ an algorithm in $\mathcal{A}_D$. For each assignment ω and the query history $(\alpha^1, \cdots, \alpha^m)$ of $(\mathbb{A}, \mathrm{tr}_i^u(\omega))$, the i-th u-transposition of $\mathbb{A}$, denote $\mathrm{tr}_i^u(\mathbb{A})$, has the query history $(\beta^1, \cdots, \beta^m)$ such that $\beta^j = \mathrm{tr}_i^u(\alpha^j)$ for each $j \leq m$.*

Note that $C(\mathbb{A}, \mathrm{tr}_i^u(\omega)) = C(\mathrm{tr}_i^u(\mathbb{A}), \omega)$.

Definition 2.4 (Equivalent assignment class, closeness, connectness). *For $\mathcal{T}_n^h$, any assignments ω, ω', we denote $\omega \approx \omega'$ if $\omega' = \mathrm{tr}_i^u(\omega)$ for some u, i. An assignment ω is equivalent to ω' if there exists a sequence of assignments $\langle \omega_i \rangle_{i=1,\cdots,s}$ such that $\omega \approx \omega_1 \approx \cdots \approx \omega_s \approx \omega'$ for some $s \in \mathbb{N}$. Then we denote $[\![\omega]\!]$ as the equivalent assignment class of ω.*

• A set Ω of assignments is closed if $\Omega = \bigcup_{\omega \in \Omega} [\![\omega]\!]$.

• A set Ω of assignments is connected if for any assignments $\omega, \omega' \in \Omega$, there exists a sequence of assignments $\langle \omega_i \rangle_{i=1,\cdots,s}$ in Ω such that $\omega \approx \omega_1 \approx \cdots \approx \omega_s \approx \omega'$.

• Given $\mathcal{A} \subseteq \mathcal{A}_D$, $\mathcal{A}$ is closed (under transposition) if for any $\mathbb{A} \in \mathcal{A}$, each internal node u and $i < n$, $\mathrm{tr}_i^u(\mathbb{A}) \in \mathcal{A}$.

Definition 2.5 (*i*-set for *n*-branching trees, adapted from[3]).
Given T_n^h, $i \in \{0, 1\}$, *i-set consists of assignments such that*
• *the root has value i,*
• *if an AND-node has value 0 (or OR-node has value 1), just one of its children has value 0 (1), and all the other $n - 1$ children have 1 (0).*

Note that *i*-set is closed and connected for $i \in \{0, 1\}$.

Definition 2.6 (*i**-set, *i'*-set). *Given* T_n^h, $i \in \{0, 1\}$,
• *i*-set is the set of all assignments ω such that $\omega(\varepsilon) = i$ and $\omega \notin i$-set.*
• *A closed set Ω of assignments is called an i'-set if it is not i-set and for any $\omega \in \Omega$, $\omega(\varepsilon) = i$.*

Definition 2.7 (*E^i*-distribution from[3]). *Suppose $\mathcal{A}$ is a subset of $\mathcal{A}_D$. A distribution d on i-set is called an E^i-distribution w.r.t. $\mathcal{A}$ if there exists $c \in \mathbb{R}$ such that for any $\mathbb{A} \in \mathcal{A}$, $C(\mathbb{A}, d) = c$.*

3. The distributional complexity for AND-OR trees

We mainly investigate the eigen distribution that achieves the distributional complexity for game trees in two kinds of probability distribution: CD and ID.

3.1. *The eigen distribution on CDs for multi-branching trees*

In,[3] Liu and Tanaka first investigate the eigen distribution for the AND-OR uniform binary tree T_h^2, and showed that the E^1-distribution is the unique eigen distribution. Suzuki and Nakamura[10] checked that the uniqueness false with respect to only directional algorithms. Peng *et al.*[7] extend their studies to balanced multi-branching trees. Okisaka *et al.*[5] generalized the results of Peng *et al.* in[7] to the case of weighted trees.

In,[3] Liu and Tanaka introduced a reverse assigning technique to formulate sets of assignments for T_2^h, namely 1-set and 0-set, in CD case.

Definition 3.1 (Reverse assigning technique in[3]). *The technique to form the 1-set (respectively, 0-set) of a tree T_2^k includes tree stages:*

(1). Assign a 1 to the root of tree T_2^k.

(2). From the root to the leaves, assign a 0 or 1 to each child of any internal node as follows

- *for an AND-node with value 1, assign 1's to all its children;*
- *for an OR-node with value 0, assign 0's to all its children;*
- *for an AND-node with value 0, assign at random a 0 to one of its children and a 1 to the other one;*
- *for the OR-node with value 1, assign at random a 1 to one of its children and a 0 to the other one.*

(3). Form the 1-set (respectively, 0-set) by collecting all the possible assignments.

Liu and Tanaka showed that for any correlated distribution d on T_2^h, the following are equivalent.

(1). d is the eigen-distribution.

(2). d is an E^1-distribution.

(3). d is the uniform distribution on the 1-set.

Suzuki and Nakamura[10] furthermore studied certain subsets of deterministic algorithms on T_2^h. Using the no-free-lunch theorem, they showed the relation of eigen-distribution and E^1- distribution with respect to "closed" subset of alpha-beta pruning algorithms:

Theorem 3.1 (Suzuki and Nakamura[10]). *Assume that T_2^h is a AND-OR tree. Suppose that d is a probability distribution on the set of assignments and $\mathcal{A}$ is any closed subset of alpha-beta pruning algorithms. Then the followings are equivalent.*

(a) d is an eigen distribution with respect to $\mathcal{A}$.

(b) d is an E^1-distribution with respect to $\mathcal{A}$.

Moreover, they proved that the uniqueness of the eigen-distribution fails if we restrict ourselves to directional algorithms, but holds for all deterministic algorithms.

Theorem 3.2 (Suzuki and Nakamura[10]). *Suppose that $h \geq 2$, where h is the height of T. Then the followings hold.*

- E^i-*distribution w.r.t. all directional algorithms is not unique. Hence, there are uncountably many eigen distribution w.r.t. all directional algorithms.*
- E^i-*distribution w.r.t. all deterministic algorithms is uniform. Thus, it is the eigen distribution w.r.t. all deterministic algorithms.*

Peng *et al.* extend the above result to balanced multi-branching case.

Theorem 3.3 (Peng *et al.*[7]). *Assume an AND-OR tree T_n^h, d is a probability distribution on the assignments, $\mathcal{A}$ is a closed subset of $\mathcal{A}_D$. Then the following two conditions are equivalent.*

(a) d is an eigen-distribution w.r.t. $\mathcal{A}$.
(b) d is an E^1-distribution w.r.t. $\mathcal{A}$.

We also consider non-directional algorithms, which play an important role to investigate the uniqueness of eigen-distribution. While a deterministic algorithm $\mathbb{A}$ works, the order of searching leaves may depend on the query history. If so, $\mathbb{A}$ is called a non-directional algorithm. We first provide an example of a such algorithm.

Example 3.1. Given a tree $\mathcal{T}_3^2$, where each leaf is labeled from left to right as shown in Fig. 3. Let $\mathbb{A}$ be a directional algorithm on $\mathcal{T}_3^2$ denoted as **123456789**, it means the algorithm evaluates the

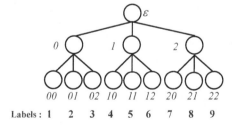

Fig. 3. $\mathcal{T}_3^2$ with label on leaves

leaves from left to right. We can define a non-directional algorithm $\mathbb{A}'$ denoted as $\widehat{1}23\underline{456789}$, where the order of searching leaves depends on the query history as follows

- if $\omega(00) = 1$, then the algorithm continues as the searching order **789456**;
- otherwise, the algorithm continues from left to right as the searching order **456789**.

Theorem 3.4. *For any tree T_n^2, E^i-distribution w.r.t. $\mathcal{A}_D$ is uniform. Thus eigen-distribution w.r.t. $\mathcal{A}_D$ is unique.*

Okisaka *et al.*[5] investigated the eigen-distribution for multi-branching trees weighted with (a, b) on correlated distributions, which is a weak version of Saks and Wigderson's (1986)[9] weighted trees.

Definition 3.2 (Okisaka *et al.*[5]). *Let A be an algorithm, ω an assignment, $\sharp_1(A, \omega)$ (resp., $\sharp_0(A, \omega)$) denote the number of leaves probed by A and assigned 1 (resp., 0) on ω. For any positive real numbers a, b,*

$$C(A, \omega; a, b) := a \cdot \sharp_1(A, \omega) + b \cdot \sharp_0(A, \omega),$$

is called a generalized cost weighted with (a, b). Obviously, $C(A, \omega) = C(A, \omega; 1, 1)$.

For a distribution d on Ω, the expected generalized cost $C(A, d; a, b) := \sum_{\omega \in \Omega} d(\omega) \cdot C(A, \omega; a, b)$. We may say that $\mathcal{T}$ is a tree weighted with (a, b) if we consider the above generalized cost. For weighted tree $\mathcal{T}$, even two leaves have the same label, they may have different weights.

Theorem 3.5 (Okisaka *et al.*[5]). *For any balanced multi-branching tree weighted with (a, b), the uniqueness of eigen-distribution holds w.r.t. all deterministic algorithms, while the uniqueness fails if we are restricted to directional algorithms.*

In,[5] Okisaka *et al.* only consider the uniqueness of eigen distribution for trees weighted with (a, b), but it is still open that what is the relation between eigen-distribution and E^i-distribution for such trees. Recently, we show the following conclusion:

Theorem 3.6. *For any trees weighted with (a, b), the eigen distribution is equivalent to the E^1-distribution with respect to the class of deterministic algorithms.*

3.2. The eigen distribution on IDs for multi-branching trees

In this part, we treat independent distributions in the context of balanced multi-branching trees. For the ID case, Liu and Tanaka proposed the following conclusion, which is called Liu-Tanaka theorem.

Theorem 3.7 (Liu and Tanaka[3]). *For any AND-OR tree T_h^2, suppose that a distribution d is an independent distribution and achieves the distributional complexity, then d is an independent identical distribution.*

They write 'it is not hard' to show the theorem and omit the proof. In,[11] Suzuki and Niida first give the following technical lemma, and furthermore showed a stronger form of the above theorem, more precisely, their results implies Liu-Tanaka theorem.

Lemma 3.1. *Given a uniform binary OR-AND tree, suppose that the distribution on T is an IID with all leaves assigned probability x and we follow the algorithm SOLVE (or equivalently any depth-first directional algorithm A). Then*

(1) $p_\sigma(x)$ is a strictly increasing function of x.

(2) $\frac{C_\sigma(x)}{p_\sigma(x)}$ is strictly decreasing.

(3) $\frac{C'_\sigma(x)}{p'_\sigma(x)}$ is strictly decreasing if σ is not a leaf.

Theorem 3.8 (Suzuki and Niida[11]). *Suppose that r is a real number such that $0 < r < 1$. Suppose that we restrict ourselves to distributions such that the probability of the root is r. Under this constraint, the statement of the above result of Liu and Tanaka still holds.*

In,[8] we define a directional algorithm DIR_d for for any $d \in ID$, and show it is optimal among all the depth-first algorithms with respect to d for any multi-branching tree. Our result can be seen as a generalization of Tarsi's theorem[13] that SOLVE is optimal for IID.

Definition 3.3. For any uniform binary tree T and $d \in ID$ on T, the depth-first directional algorithm DIR_d is defined inductively as follows. The basic case is trivial. For the induction case, let $\sigma * i$ ($i = 1, 2$) be a child of non-terminal node σ, and assume $\mathrm{DIR}_{d_{\sigma * i}}$ has been defined for each subtree $T_{\sigma * i}$.

(1) In the case that σ is labeled $\wedge$, DIR_{d_σ} is the concatenation of $\mathrm{DIR}_{d_{\sigma * 1}}$ and $\mathrm{DIR}_{d_{\sigma * 2}}$ (denote $\mathrm{DIR}_{d_\sigma} := \mathrm{DIR}_{d_{\sigma * 1}} \cdot \mathrm{DIR}_{d_{\sigma * 2}}$) if

$$\frac{C_{\sigma * 1}(\mathrm{DIR}_{d_{\sigma * 1}}, d_{\sigma * 1})}{q_{\sigma * 1}} \leq \frac{C_{\sigma * 2}(\mathrm{DIR}_{d_{\sigma * 2}}, d_{\sigma * 2})}{q_{\sigma * 2}},$$

otherwise $\mathrm{DIR}_{d_\sigma} := \mathrm{DIR}_{d_{\sigma * 2}} \cdot \mathrm{DIR}_{d_{\sigma * 1}}$.

(2) In the case that σ is labeled $\vee$, $\mathrm{DIR}_{d_\sigma} := \mathrm{DIR}_{d_{\sigma * 1}} \cdot \mathrm{DIR}_{d_{\sigma * 2}}$ if

$$\frac{C_{\sigma * 1}(\mathrm{DIR}_{d_{\sigma * 1}}, d_{\sigma * 1})}{1 - q_{\sigma * 1}} \leq \frac{C_{\sigma * 2}(\mathrm{DIR}_{d_{\sigma * 2}}, d_{\sigma * 2})}{1 - q_{\sigma * 2}},$$

otherwise $\mathrm{DIR}_{d_\sigma} := \mathrm{DIR}_{d_{\sigma * 2}} \cdot \mathrm{DIR}_{d_{\sigma * 1}}$.

Peng *et al.* showed the following two theorems in.[8]

Theorem 3.9 (Peng *et al.*[8]). *For any multi-branching tree, if $d \in ID$, DIR_d is optimal among all depth first algorithms.*

Proof. We prove this by induction on height h. The base case is trivial. For the induction step, let T be a uniform binary tree with height $h + 1$, where the root λ is labeled $\wedge$. The other case can be shown similarly.

Suppose that DIR_{d_i} is optimal for each subtree T_i with height h. Let Ω_{h+1} be the set of assignments for T, Ω_h and Ω'_h the set of assignments for T_1 and T_2. For any $d \in ID$ on T, there exist d_i for

$T_i (i = 1, 2)$ such that $d(\omega) = d_1(\omega_1) \times d_2(\omega_2)$, where $\omega = \omega_1 \omega_2$, $\omega_1 \in \Omega_h$ and $\omega_2 \in \Omega'_h$. For any depth-first algorithm A and $d \in \mathrm{ID}$, if A evaluates the subtree T_1 first, then $C(A, d) = \sum_{\omega \in \Omega_{h+1}} C(A, \omega) \cdot$
$d(\omega) = \sum_{\omega \in \Omega^0} C(A, \omega) \cdot d(\omega) + \sum_{\omega \in \Omega^1} C(A, \omega) \cdot d(\omega)$, where $\Omega^i := \{\omega \in \Omega_{h+1} \mid$ the root of T_1 has value i with $\omega\}$.

Assume A is a depth-first non-directional algorithm. By A_1, we denote the algorithm of A for T_1, and by A_{ω_1} the algorithm of A for T_2 depending on the assignment ω_1 in T_1. Thus, $C(A, d) = \sum_{\omega_1 \in \Omega_h^0} C_{\lambda *1}(A_1, \omega_1) \cdot d_1(\omega_1) + \sum_{\omega_1 \in \Omega_h^1} \sum_{\omega_2 \in \Omega'_h} (C_{\lambda *1}(A_1, \omega_1) +$
$C_{\lambda *2}(A_{\omega_1}, \omega_2)) \cdot d_2(\omega_2) d_1(\omega_1)$, where $\Omega_h^i := \{\omega_1 \in \Omega_h \mid$ the root of T_1 has value i with $\omega_1\}$. We calculate that $C(A, d) = C_{\lambda *1}(A_1, d_1) +$
$\sum_{\omega_1 \in \Omega_h^1} d_1(\omega_1) \cdot C_{\lambda *2}(A_{\omega_1}, d_2)$. By induction hypothesis, we can take algorithms DIR_{d_1} and DIR_{d_2} such that $C(A, d) \geq C_{\lambda *1}(\mathrm{DIR}_{d_1}, d_1) +$
$(1 - q_1) C_{\lambda *2}(\mathrm{DIR}_{d_2}, d_2)$. Let $A'_1 = \mathrm{DIR}_{d_1} \cdot \mathrm{DIR}_{d_2}$, then clearly $C(A, d) \geq C(A'_1, d)$.

Using the similar arguments as above, if A evaluates T_2 first, we can get algorithm $A'_2 = \mathrm{DIR}_{d_2} \cdot \mathrm{DIR}_{d_1}$. Thus, $C(A, d) \geq C(A'_2, d)$. If $\frac{C_{\lambda *1}(\mathrm{DIR}_{d_1}, d_1)}{q_1} \leq \frac{C_{\lambda *2}(\mathrm{DIR}_{d_2}, d_2)}{q_2}$, then $C(A'_1, d) = C(A'_2, d) -$
$(q_1 C_{\lambda *2}(\mathrm{DIR}_{d_2}, d_2) - q_2 C_{\lambda *1}(\mathrm{DIR}_{d_1}, d_1)) \leq C(A'_2, d)$. By Definition 3.3, DIR_d is A'_1 if $\frac{C_{\lambda *1}(A_{d_1}, d_1)}{q_1} \leq \frac{C_{\lambda *2}(A_{d_2}, d_2)}{q_2}$, otherwise A'_2. Clearly, DIR_d is optimal among all the depth-first algorithms. $\square$

Let $\mathrm{ID}(r)$ denote the set of independent distributions which induce that the probability of the root having value 0 is r. We prove a technical lemma which is a generalization of "fundamental relationships between costs and probabilities" due to Suzuki and Niida in,[11] and show the following.

Theorem 3.10 (Peng et al.[8]). *For any balanced multi-branching AND-OR tree $\mathcal{T}$, we fix $\delta \in \mathrm{ID}(r)$ and $0 < r < 1$. If the following equation holds,*

$$\min_{A:\text{depth}} C(A, \delta) = \max_{d \in \mathrm{ID}(r)} \min_{A:\text{depth}} C(A, d),$$

then $\delta \in \mathrm{IID}$.

In the above Theorem 3.10, by "A : depth", we emphasize "A ranges over the depth-first algorithms". Suzuki[12] extended Theorem 3.10 to a wider class of deterministic algorithms which also includes non-depth-first ones.

Theorem 3.11 (Suzuki[12]). *Suppose that r is a real number such that $0 < r < 1$. For any balanced multi-branching AND-OR tree, suppose d_1 satisfies the following equation,*

$$\min_{A:all} C(A, d_1) = \max_{d \in ID(r)} \min_{A:all} C(A, d).$$

Then, there exists a depth-first directional algorithm B_0 such that $C(B_0, d_1) = \min_{A:all} C(A, d_1)$. In addition, d_1 is an IID.

Okisaka *et al.*[5] investigated the eigen-distribution for multi-branching trees weighted with (a, b) on correlated distributions. In,[15] we concentrate on the studies of eigen-distribution for multi-branching weighted trees on independent distributions. In particular, we generalize our previous results in Peng *et al.* (2017)[8] to weighted trees where the cost of querying each leaf is associated with the leaf and its Boolean value. For a multi-branching weighted tree, we define a directional algorithm and show it is optimal among all the depth-first algorithms with respect to the given independent distribution.

Definition 3.4 (Weighted tree). *Given $d \in$ ID and a directional algorithm A on $\mathcal{T}$ with weights w, for each node σ, we inductively define $C_\sigma^i(A, d; w)$, or simply C_σ^i to be the expected cost when σ gets value $i = 0, 1$, and $p_\sigma(d)$ the probability of σ being 0 as follows.*

(1) *If σ is the n-th (from left to right) leaf, then $p_\sigma(d) := p_n$ and $C_\sigma^i := w_i(n)$ for $i = 0, 1$.*

(2) *Suppose σ is a non-terminal node with children $\sigma_1, \ldots, \sigma_n$ and algorithm A evaluates the children $\sigma_{f(1)}, \ldots, \sigma_{f(n)}$ in this order, where f is a permutation of $\{1, \ldots, n\}$. For simplicity, we write $p_{\sigma_k}(d)$ as p_{σ_k}. If σ is labelled $\wedge$, then*

$$p_\sigma(d) = 1 - \prod_{k=1}^{n} (1 - p_{\sigma_k})$$

and the expected costs to evaluate σ as value i are

$$C_\sigma^1 := \prod_{k=1}^{n} (1 - p_{\sigma_{f(k)}}) \cdot \left(\sum_{j=1}^{n} C_{\sigma_{f(j)}}\right),$$

$$C_\sigma^0 = p_{\sigma_{f(1)}} \cdot C_{\sigma_{f(1)}} + \sum_{l=2}^{n} \left(p_{\sigma_{f(l)}} \prod_{k=1}^{l-1} (1 - p_{\sigma_{f(k)}}) \cdot \left(C_{\sigma_{f(l)}} + \sum_{j=1}^{l-1} C_{\sigma_{f(j)}}\right) \right),$$

$$C_\sigma := C_\sigma^0 + C_\sigma^1.$$

If σ is labelled $\vee$, they are defined analogously.

n parallel to Definition 3.4, we will construct a directional algorithm DIR_d w.r.t. a $d \in \mathsf{ID}$, based on a greedy strategy to minimize the expected cost.

Definition 3.5. Given $d \in \mathsf{ID}$ on $\mathcal{T}$ with weights w, we inductively define a directional algorithm DIR_d as follows. For each node σ of $\mathcal{T}$,

(1) In the case that σ is labeled $\wedge$, for the lexicographically minimal permutation f on $\{1, \cdots, n\}$ such that

$$\frac{C_{\sigma_{f(1)}}(\mathsf{DIR}_{d_{\sigma_{f(1)}}}, d_{\sigma_{f(1)}}; w)}{p_{\sigma_{f(1)}}} \leq \cdots \leq \frac{C_{\sigma_{f(n)}}(\mathsf{DIR}_{d_{\sigma_{f(n)}}}, d_{\sigma_{f(n)}}; w)}{p_{\sigma_{f(n)}}},$$

we set $\mathsf{DIR}_d := \mathsf{DIR}_{d_{f(1)}} \cdots \mathsf{DIR}_{d_{f(n)}}$, it is executed from left to right.

(2) In the case that σ is labeled $\vee$, for the lexicographically minimal permutation f on $\{1, \cdots, n\}$ such that

$$\frac{C_{\sigma_{f(1)}}(\mathsf{DIR}_{d_{\sigma_{f(1)}}}, d_{\sigma_{f(1)}}; w)}{1 - p_{\sigma_{f(1)}}} \leq \cdots \leq \frac{C_{\sigma_{f(n)}}(\mathsf{DIR}_{d_{\sigma_{f(n)}}}, d_{\sigma_{f(n)}}; w)}{1 - p_{\sigma_{f(n)}}},$$

we set $\mathsf{DIR}_d := \mathsf{DIR}_{d_{f(n)}} \cdots \mathsf{DIR}_{d_{f(1)}}$, it is also executed from left to right.

The following example show how to calculate the expected cost for a multi-branching weighted tree and construct the directional algorithm DIR_d by using Definitions 3.4 and 3.5.

Example 3.2. Let $A_1 = 12$ and $A_2 = 21$ be two directional algorithms on a weighted tree $\mathcal{T}_2^1$, and $d \in \mathrm{ID}$. Here, A_1 first queries the leftmost leaf of $\mathcal{T}_2^1$ (A_2 first queries the rightmost leaf of $\mathcal{T}_2^1$, respectively), then algorithm reads another one. If the root λ is labeled $\wedge$, $C_\lambda(A_1, d; w) = p_1 w_0(1) + (1 - p_1)p_2(w_1(1) + w_0(2)) + (1 - p_1)(1 - p_2)(w_1(1) + w_1(2))$, and $C_\lambda(A_2, d; w) = p_2 w_0(2) + (1 - p_2)p_1(w_1(2) + w_0(1)) + (1 - p_2)(1 - p_1)(w_1(2) + w_1(1))$. By a simple calculation, $C_\lambda(A_1, d; w) \le C_\lambda(A_2, d; w)$ iff

$$\frac{p_1 w_0(1) + (1 - p_1)w_1(1)}{p_1} \le \frac{p_2 w_0(2) + (1 - p_2)w_1(2)}{p_2}$$

iff $\mathrm{DIR}_d = A_1$ (or possibly A_2 if the equality hold).

By induction, we show the following.

Theorem 3.12. *For any weighted tree and $d \in \mathrm{ID}$, DIR_d is optimal among all the depth-first algorithms.*

A tree $\mathcal{T}_n^h$ is called weighted with (a, b) if $w_0 = b$ and $w_1 = a$ for all leaves. Thus, Theorem 3 in[8] can be generalized as follows.

Theorem 3.13. *For any balanced multi-branching AND-OR tree $\mathcal{T}$ weighted with (a, b), suppose that $\delta \in \mathrm{ID}(r)$ and $0 < r < 1$. If the following equation holds,*

$$\min_{A:\text{depth}} C(A, \delta, a, b) = \max_{d \in \mathrm{ID}(r)} \min_{A:\text{depth}} C(A, d, a, b),$$

then $\delta \in \mathrm{IID}$.

Proof. It is enough to show that for any $d \in \mathrm{ID}$ on $\mathcal{T}$, there exists $d' \in \mathrm{IID}$ such that $C(\mathrm{DIR}_d, d, a, b) \le C(\mathrm{DIR}_d, d', a, b)$ and the probability at the root σ, $p_\sigma(d') = p_\sigma(d)$ ($p_\sigma(d) \ne 0$ or 1).

We prove this by induction on height h. In the case $h = 1$, without loss of generality, we assume the given d satisfying $p_1 \le p_2 \le \cdots \le p_i \le p_{i+1} \le \cdots \le p_n$. We show that we can adjust the values of p_i and p_{i+1} such that the cost strictly increases while all numbers other than p_i and p_{i+1} are fixed. We may denote the cost and probability by $C(p_i, p_{i+1})$ and $p_{i,i+1}$ respectively.

For simplicity, we only consider the case where the root is labeled $\vee$. Here DIR_d is SOLVE.[9] Following DIR_d, we have $C(p_1, p_2) = m + n(C(p_i) + p_i(C(p_{i+1}) + p_{i+1}\mu))$, where $C(p_i) = p_i b + (1 - p_i)a$, $C(p_{i+1}) = p_{i+1}b + (1 - p_{i+1})a$, m, n, μ are constants with $\mu > 0$; and $p_{i,i+1} = p_i p_{i+1}$. Since we keep $p_\sigma(d') = p_\sigma(d)$, $p_{i,i+1}$ is a constant, denote $p_{i,i+1} = c$. Thus, $C'(p_i(p_{i+1}), p_{i+1}) = (C(p_i) + p_i(C(p_{i+1}) + p_{i+1}\mu)))' = -\frac{cb}{p_{i+1}^2}$. Since $C'(p_i(p_{i+1}), p_{i+1}) < 0$ and $p_{i,i+1} = c$, the cost $C(p_i, p_{i+1})$ will increase if $p_i \neq p_{i+1}$.

Since Lemma 3.1 still holds for the general case ($h \geq 2$) of balanced multi-branching trees weighted with (a, b), we can calculate the $C' < 0$ in case from segment-wise IID to IID. □

Recently, we show the following conclusion which is a generalization of Suzuki and Niida's result in[11] to multi-branching cases.

Lemma 3.2. *Suppose that T_n^h is an AND-OR tree or an OR-AND tree. Let $i \in \{0, 1\}$, d_0 be an ID such that each leaf has the probability $1 - i$. Then, in the case that h is even, denoted by $h = 2k$, $\min_A C(A, d_0) = n^k$. In the case that h is odd, denoted by $2k + 1$, the value of $\min_A C(A, d_0)$ is as follows:*

- $\min_A C(A, d_0) = n^k$ *if we consider AND-OR tree and $i = 0$, or we consider OR-AND tree and $i = 1$.*
- $\min_A C(A, d_0) = n^{k+1}$, *otherwise.*

We show that the independent distribution is identical when the distributional complexity holds without considering the restriction for the probability of the root, this result still holds for AND-OR balanced trees weighted with (a, b).

4. Future research

In the future, we would like to investigate the eigen-distribution of weighted game trees with different weights on different leaves, or MIN-MAX trees with real values. In fact, such studies have been already carried out by many researchers, e.g., Greiner *et al.*[1]

References

1. Greiner R, Hayward R, Jankowska M, Molloy M (2006) Finding optimal satisficing strategies for and-or trees. Artif Intell 170(1):19-58
2. Knuth DE, Moore, RW (1975) An analysis of alpha-beta pruning. Artif Intell 6(4):293-326
3. Liu CG, Tanaka K (2007) Eigen-distribution on random assignments for game trees. Inform Process Lett 104(2):73-77
4. Liu CG, Tanaka K (2007) The computational complexity of game trees by eigen-distribution. In: Proc. 1st International Conference on COCOA. Springer, pp. 323-334
5. Okisaka S, Peng W, Li W, Tanaka K (2017) The eigen-distribution of weighted game trees. In: Proc. 11th Annual International Conference on COCOA. Springer, pp. 286-297
6. Pearl J (1980) Asymptotic properties of minimax trees and game-searching procedures. Artif Intell 14(2):113-138
7. Peng W, Okisaka S, Li W, Tanaka K (2016) The uniqueness of eigen-distribution under non-directional algorithms. IAENG Int J Comput Sci 43(3):318-325
8. Peng W, Peng N, Ng K, Tanaka K, Yang Y (2017) Optimal depth-first algorithms and equilibria of independent distributions on multi-branching trees. Inform Process Lett 125:41-45
9. Saks M, Wigderson A (1986) Probabilistic Boolean decision trees and the complexity of evaluating game trees. In: Proceeding of 27th Annual IEEE Symposium on FOCS. Springer, pp. 29-38
10. Suzuki T, Nakamura R (2012) The eigen distribution of an AND-OR tree under directional algorithms. IAENG Int J Appl Math 42(2):122-128
11. Suzuki T, Niida Y (2015) Equilibrium points of an AND-OR tree: under constraints on probability. Ann Pure Appl Logic 166(11):1150-1164
12. Suzuki T (2018) Non-depth-first search against independent distributions on an AND-OR tree. Inform Process Lett 139:13-17
13. Tarsi M (1983) Optimal search on some game trees. J ACM 30(3):389-396
14. Yao A.C.C (1977) Probabilistic computations: toward a unified measure of complexity. In: Proceeding 18th Annual IEEE Symposium on FOCS. Springer, pp. 222-227
15. Peng W, Peng N, Tanaka K (2021) Optimal search on some game trees. Methodology and Computing in Applied Probability 24:277-287

Rational Sequences Converging to Left-c.e. Reals of Positive Effective Hausdorff Dimension

Hiroyuki Imai[1], Masahiro Kumabe[2], Kenshi Miyabe[3], Yuki Mizusawa[4] and Toshio Suzuki[5,*]

[1,4,5] *Department of Mathematical Sciences, Tokyo Metropolitan University, Hachioji, Tokyo 192-0397, Japan*
[5] *E-mail: toshio-suzuki@tmu.ac.jp*

[2] *Faculty of Liberal Arts, The Open University of Japan, Mihama-ku, Chiba 261-8586, Japan*
E-mail: kumabe@ouj.ac.jp

[3] *School of Science and Technology, Meiji University, Kawasaki, Kanagawa 214-8571, Japan*
E-mail: research@kenshi.miyabe.name

In our previous work, we characterized Solovay reducibility using Lipschitz condition, and introduced quasi Solovay reducibility (qS-reducibility, for short) as a Hölder condition counterpart. In this paper, we investigate effective dimensions and ideals closely related to quasi Solovay reducibility by means of the rate of convergence. We show that the qS-completeness among left-c.e. reals is equivalent to having a positive effective Hausdorff dimension. The Solovay degrees of qS-complete left-c.e. reals form a filter. On the other hand, the Solovay degrees of non-qS-complete left-c.e. reals do not form an ideal. Based on observations on the relationships between rational sequences and reducibility, we introduce a stronger version of qS-reducibility. Given a degree of this reducibility, the lower cone (including the given degree) forms an ideal. By developing these investigations, we characterize the effective dimensions by means of the rate of convergence. We give a variation of the first incompleteness theorem based on Solovay reducibility.

Keywords: Solovay Reducibility; Quasi Solovay Reducibility; Effective Dimension; Ideal; Rate of Convergence.

2020 Mathematics Subject Classification: 03D32, 68Q30

---

*The corresponding author. Partially supported by JSPS KAKENHI Grant Number JP16K05255.

1. Introduction

1.1. *Background*

Martin-Löf randomness, also called 1-randomness, has been the central notion in the theory of algorithmic randomness. Informally speaking, a real is random if its binary expansion is complicated in a certain algorithmic sense. In particular, left-c.e. random reals have many interesting properties. Here, a real number is called *left-c.e.* if the left set of its Dedekind cut is computably enumerable. For the general background on the study of randomness, see Downey and Hirschfeldt[6] or Nies[13].

The starting point of our discussion is the following theorem.

Theorem 1.1. (Demuth[5], Downey *et al.*[7]). *If the sum of two left-c.e. reals α, β is 1-random, then at least one of α or β is 1-random.*

See also Corollary 9.5.9 of Ref. 6. The theorem intuitively says that, in the unit interval of the real line, the addition of two non-random reals results in a non-random real.

Our main goal of this paper is to reinvestigate the theorem above in terms of the rate of the convergence inspired by the recent result by Barmpalias and Lewis (Theorem 1.2) and our previous work on quasi Solovay reducibility.

Solovay reducibility is a preorder that compares two reals in terms of algorithmic complexity. A real α is *Solovay reducible* to a real β, $\alpha \leq_S \beta$ in symbols, if there exists a partial computable function f from $\mathbb{Q}$ to $\mathbb{Q}$ and a positive constant c such that for each rational $q < \beta$, $f(q) < \alpha$ is defined and $\alpha - f(q) < c(\beta - q)$. Informally speaking, if we have a good approximation q of β then we get a good approximation $f(q)$ of α. If we restrict ourselves to the left-c.e. reals, Solovay completeness coincides with 1-randomness. Miyabe *et al.*[12] studied effective dimensions and Solovay degrees. Among others, the sum of non-random reals is studied by means of the concept of ideals of the partially ordered set.

We may view Solovay reducibility from the perspective of analysis. In our previous work (Kumabe *et al.*[8]), we showed that for left-c.e. reals α and β, $\alpha \leq_S \beta$ if and only if there exists a Lipschitz continuous function from $(-\infty, \beta)$ to $(-\infty, \alpha)$ satisfying certain conditions. We also introduced the concept of *quasi Solovay reducibility*, which corresponds to the Hölder continuous functions. A real α is quasi Solovay reducible to a real β, $\alpha \leq_{qS} \beta$ in symbols, if there exists a partial computable function f from $\mathbb{Q}$ to $\mathbb{Q}$ and positive constants d, ℓ such that for each rational $q < \beta$, $f(q) < \alpha$ is defined and $(\alpha - f(q))^{\ell} < d(\beta - q)$. We showed that for left-c.e. reals α and β, $\alpha \leq_{qS} \beta$ if and only if there exists a Hölder continuous function from $(-\infty, \beta)$ to $(-\infty, \alpha)$ satisfying certain conditions.

In this paper, we investigate effective dimensions and ideals closely related to quasi Solovay reducibility. The most important concept in our method is the rate of convergence. Barmpalias and Lewis-Pye[2] investigate a quantity similar to the left-hand derivative.

Theorem 1.2. (Barmpalias and Lewis-Pye[2]. See also Miller[11]). *Suppose $\langle a_n \rangle \nearrow \alpha$, $\langle b_n \rangle \nearrow \beta$. If β is random, then the following hold.*

$$\lim_{n \to \infty} \frac{\alpha - a_n}{\beta - b_n} \quad exists.$$

Moreover,

- *The limit value is independent of the choice of sequences.*
- *The limit value = 0 if and only if α is not 1-random.*

1.2. *Overview*

In Section 2, we characterize qS-completeness among left-c.e. reals by means of Solovay reducibility and dimension. In particular, qS-completeness is equivalent to having a positive dimension. In Section 3, we investigate some ideals of left-c.e. reals. We show that the Solovay degrees of qS-complete reals form a filter. On the other hand, the Solovay degrees of non-qS-complete reals do not form an ideal. We investigate the relationships between rational sequences and reducibility, and by means of those investigations, we introduce a stronger version of qS-reducibility. In the stronger version, the lower cone below ($\leq$) a given degree forms an ideal. In Section 4, by developing Section 3, we characterize the effective dimensions by means of the rate of convergence. Section 5 is an appendix. As a by-product of our observation, we show a variation of the first incompleteness theorem by means of Solovay reducibility.

1.3. *Notation*

In this paper, $\langle a_n \rangle_{n \in \mathbb{N}}$, or $\langle a_n \rangle$ denotes a sequence of numbers. Unless otherwise specified, α and β denote left-c.e. reals.

Convention on sequences: Throughout the paper, unless otherwise specified, $\langle a_n \rangle \nearrow \alpha$ denotes that $\langle a_n \rangle$ is a strictly increasing computable sequence of rationals, and $\langle a_n \rangle$ converges to α.

We are mainly interested in Solovay reducibility and quasi Solovay reducibility and their stronger versions. In their analysis, K-reducibility and strong K-reducibility play an important role to connect our analysis to effective dimensions.

Definition 1.1. A real α is called *K-reducible* to a real β, $\alpha \leq_K \beta$ in symbols, if $K(\alpha \restriction n) \leq K(\beta \restriction n) + O(1)$. Here, K is the prefix-free Kolmogorov complexity , and $\alpha \restriction n$ is the first n bits of the binary expansion of α.

Definition 1.2. A real α is *strongly K-reducible* to a real β, $\alpha \ll_K \beta$ in symbols, if $\lim_{n \to \infty} (K(\beta \restriction n) - K(\alpha \restriction n)) = \infty$.

Clearly, if $\alpha \ll_K \beta$, then $\alpha \leq_K \beta$. The converse does not hold; we will see stronger results later.

2. Characterization of qS-complete reals

In this section, we characterize qS-complete reals by means of Solovay reducibility and dimension. In the first subsection, we review previous work on Solovay reducibility and K-reducibility, and we show a slightly stronger result than the original. Throughout the paper, we sometimes use Lemma 2.1. In the second subsection, we show that qS-complete left-c.e. reals are characterized by means of a positive effective Hausdorff dimension.

2.1. *Solovay reducibility and K-reducibility*

Solovay reducibility implies K-reducibility. It is known that the inverse implication does not hold (Ref. 7). On the other hand, strong K-reducibility implies (a property slightly stronger than) Solovay reducibility.

Proposition 2.1. *(Prop. 2.3 of Ref. 12) Let α, β be left-c.e. reals. If $\alpha \ll_K \beta$ then $\alpha <_S \beta$.*

The proof of Prop. 2.1 in Ref. 12 is useful when we are interested in rational sequences converging to α and β under assumption of $\alpha \ll_K \beta$. For later convenience, we reconstitute the main part of the proof in a generalized form. See the notation section for the convention on sequences $\langle a_s \rangle$.

Lemma 2.1. *Let α, β be left-c.e. reals. Suppose that $\langle a_s \rangle \nearrow \alpha$ and $\langle b_s \rangle \nearrow \beta$. Let ℓ, m be positive reals.*

(1) *If $\lim_n (K(\beta \restriction \ell n) - K(\alpha \restriction n)) = \infty$, then there exists a strictly increasing sequence $\langle s_n \rangle$ of natural numbers with the following properties: For almost all n and for each s such that $s_n \leq s \leq s_{n+1}$, the inequalities Eq. (1) and Eq. (2) below hold:*

$$\alpha - a_s \leq 2^{-n} \tag{1}$$

$$\beta - b_s \geq m2^{-\ell(n+1)} \tag{2}$$

(2) *If $\limsup_n (K(\beta \restriction \ell n) - K(\alpha \restriction n)) = \infty$ then there exists a strictly increasing sequence $\langle s_n \rangle$ of natural numbers with the following properties: For infinitely many n, the above-mentioned inequalities Eq. (1) and Eq. (2) hold for all s with $s_n \leq s \leq s_{n+1}$.*

Remarks (for both assertions). (a) In the case where β is rational, the assumptions on the limit do not hold. Thus we may assume that β is irrational.

(b) The sequence $\langle s_n \rangle$ need not be computable.

(c) For simplicity, we concentrate on the case where ℓ, m are positive integers. The following proof works for the general case with minor changes. In the general case, a real number expressing the

length of a string should be replaced by a certain integer. For exam-
ple, $\beta \upharpoonright \ell n$ should be replaced by $\beta \upharpoonright \lfloor \ell n \rfloor$.

Proof. (1) Given n, let s_n be the least s such that $a_s \upharpoonright n = \alpha \upharpoonright n$. Then we have $\alpha - a_s \le \alpha - a_{s_n} \le 2^{-n}$ for each s such that $s_n \le s \le s_{n+1}$. Therefore, we have Eq. (1).

If the string $a_{s_n} \upharpoonright n$ is given, we know n as its length, and we can find s_n by means of sequence $\langle a_s \rangle$. Then we can compute $b_{s_n} \upharpoonright \ell n$, which implies the following inequality:

$$K(b_{s_n} \upharpoonright \ell n) \le K(a_{s_n} \upharpoonright n) + O(1) = K(\alpha \upharpoonright n) + O(1) \qquad (3)$$

We are going to show that for almost all n and for each nonnegative integer k such that $k \le m/2 + 1$, $\beta \upharpoonright \ell n$ is neither lexicographic kth successor of $b_{s_n} \upharpoonright \ell n$ nor lexicographic kth predecessor of $\beta \upharpoonright \ell n$, where the 0th successor (predecessor) denotes $b_{s_n} \upharpoonright \ell n$ itself. The proof is as follows. If the above-mentioned assertion fails, for infinitely many n, we have $K(\beta \upharpoonright \ell n) \le K(b_{s_n} \upharpoonright \ell n) + O(1)$. By Eq. (3), we have $K(\beta \upharpoonright \ell n) \le K(\alpha \upharpoonright n) + O(1)$, which contradicts to our assumption of $\liminf_n (K(\beta \upharpoonright \ell n) - K(\alpha \upharpoonright n)) = \infty$.

Hence, for almost all n, it holds that $\beta - b_{s_n} \ge m2^{-\ell n}$. Thus, for almost all n and for each s such that $s_n \le s \le s_{n+1}$, it holds that $\beta - b_s \ge \beta - b_{s_{n+1}} \ge m2^{-\ell(n+1)}$. Hence, for almost all n we have Eq. (2).

(2) We define $\langle s_n \rangle$ in the same way as above. Then, we have Eq. (1). In addition, for all n we have Eq. (3).

Then we can show that for infinitely many n and for each nonnegative integer k such that $k \le m/2 + 1$, $\beta \upharpoonright \ell n$ is neither lexicographic kth successor of $b_{s_n} \upharpoonright \ell n$ nor lexicographic kth predecessor of $\beta \upharpoonright \ell n$. The proof is given by means of our assumption of $\limsup_n (K(\beta \upharpoonright \ell n) - K(\alpha \upharpoonright n)) = \infty$.

Therefore, for infinitely many n, it holds that $\beta - b_{s_n} \ge m2^{-\ell n}$. Thus, for infinitely many n and for each s such that $s_n \le s \le s_{n+1}$, it holds that $\beta - b_s \ge \beta - b_{s_{n+1}} \ge m2^{-\ell(n+1)}$. Hence, for infinitely many n we have Eq. (2). $\square$

Proof. (of Proposition 2.1, sketch) Under the assumption of Lemma 2.1 assertion (1), let f be a partial function from $\mathbb{Q}$ to $\mathbb{Q}$

such that for each q, $f(q) = a_s$, where s is the least one such that $q < b_s$. Then it holds that $(\alpha - f(q))^\ell \leq (2^\ell/m)(\beta - q)$. We look at the case where $\ell = m = 1$. Thus we know: If α, β are left-c.e. and irrational then $\alpha \ll_K \beta \implies \alpha \leq_S \beta$. If $\beta \leq_S \alpha$, then $\beta \leq_K \alpha$, which contradicts $\alpha \ll_K \beta$. $\qquad\square$

2.2. Characterization of qS-complete reals by dimension

Effective Hausdorff dimension and effective packing dimension are characterized by prefix-free Kolmogorov complexity.

Theorem 2.1. (Mayordomo[10]). *Given* $\alpha \in 2^\omega$, *the effective Hausdorff dimension* $\dim(\alpha)$ *is characterized as follows.*

$$\dim(\alpha) = \liminf_n \frac{K(\alpha \restriction n)}{n} \tag{4}$$

Theorem 2.2. (Athreya, Hichcock, Lutz and Mayordomo[1]). *Given* $\alpha \in 2^\omega$, *the effective packing dimension* $\mathrm{Dim}(\alpha)$ *is characterized as follows.*

$$\mathrm{Dim}(\alpha) = \limsup_n \frac{K(\alpha \restriction n)}{n} \tag{5}$$

A survey of algorithmic dimensions may be found in Chapter 13 of Ref. 6.

For a left-c.e. real, Solovay completeness and 1-randomness are equivalent. In this section, we show that qS-completeness and $\dim(\alpha) > 0$ are equivalent. We are interested in the structure of the Solovay degrees of qS-complete sets. We will see that the set of these degrees forms a filter.

Tadaki[15], in his study on partial randomness, introduced the generalized halting probability $\Omega^T = \sum_{p\in\mathrm{dom}(U)} 2^{-|p|/T}$ for each positive real number $T \leq 1$, where p runs over the domain of a fixed universal prefix-free machine. In our previous work[8], we introduced a variation of the generalized halting probability for T of the form 2^{-n}.

Definition 2.1. Ω_{2^0} denotes Ω. For each positive integer n, letting $0.a_0a_1a_2\ldots$ be the binary expansion of $\Omega_{2^{-n}}$, we define $\Omega_{2^{-(n+1)}}$ as $0.b_0b_1b_2\ldots$, where $b_{2n} = a_n$, and $b_{2n+1} = 1 - a_n$ for each n.

In our previous work[8], we showed that $\Omega_{2^{-n}}$ are qS-complete. Now we improve the result as follows.

Theorem 2.3. *The following are equivalent for a left-c.e. real α.*

(1) *α is qS-complete.*
(2) *For some $n \in \mathbb{N}^+$, letting $T = 2^{-n}$, $\Omega_T \leq_S \alpha$.*
(3) $\dim(\alpha) > 0$

Proof. (1) $\Longrightarrow$ (3): Since $\Omega \leq_{qS} \alpha$, there are $\ell \in \mathbb{N}$ and sequences $\langle \omega_s \rangle \nearrow \Omega$ and $\langle a_s \rangle \nearrow \alpha$ such that $(\Omega - \omega_s)^{\ell} < d(\alpha - a_s)$. Let U be a universal prefix-free machine (see Section 3.5 of Ref. 6). Take a string σ such that $U(\sigma) = \alpha \upharpoonright (n\ell)$ and $|\sigma| = K(\alpha \upharpoonright (n\ell))$. We can compute ω_s such that $\Omega - \omega_s < d^{1/\ell} 2^{-n}$ by means of σ and constant bits. If $\Omega - \omega_s < 2^{-m}$ then $\Omega \upharpoonright m$ is either $\omega_s \upharpoonright m$ itself or that plus 2^{-m}, and the latter is given by additional 1 bit information. To sum up, we can find $\Omega \upharpoonright n$ by means of σ and constant bit. Therefore, it holds that $K(\alpha \upharpoonright (n\ell)) \geq K(\Omega \upharpoonright n) - O(1) \geq n - O(1)$. Hence $\dim(\alpha) \geq 1/\ell > 0$.

(3) $\Longrightarrow$ (2): For some T of the form $T = 2^{-n}$, it holds that $\mathrm{Dim}(\Omega_T) < \dim(\alpha)$. Therefore, $\Omega_T \ll_K \alpha$. By Proposition 2.1, we have $\Omega_T \leq_S \alpha$.

(2) $\Longrightarrow$ (1): By Ref. 8, $\Omega \leq_{qS} \Omega_T$. $\square$

3. Ideals on left-c.e. reals

The underlying intuition of this section is that the addition of two non-random left-c.e. reals results in a non-random real. It is known that this is the case if randomness means 1-randomness. This means the non-complete Solovay degrees form an ideal. We will investigate some ideals and filters of qS-degrees. In the first subsection, we show that the Solovay degrees of qS-complete reals form a filter. On the other hand, Solovay degrees of non-qS-complete reals do not form an ideal. In the second subsection, we observe the relationships between rational sequences and degrees. Based on these observations, in the third subsection, we introduce a stronger version of qS-reducibility that has a nice property with respect to addition.

3.1. *The filter of the Solovay degrees of qS-complete reals*

A subset F of a partially ordered set $(P, \leq_P)$ is a *filter* if the following hold.

- $a \in F \wedge a \leq_P b \implies b \in F$
- $a, b \in F \implies \exists c \in F \; c \leq_P a \wedge c \leq_P b$

A subset X is an *ideal* if the following hold.

- $a \in X \wedge b \leq_P a \implies b \in X$
- $a, b \in X \implies \exists c \in X \; a \leq_P c \wedge b \leq_P c$

We consider the Solovay degrees of left-c.e. reals. Given a Solovay degree $\mathbf{a}$, we define $\dim \mathbf{a}$ as to be $\dim(a)$ for some $a \in \mathbf{a}$. Since Solovay reducibility implies K-reducibility, this definition is well-defined (See [12] Section 4). The Solovay complete degree forms a filter obviously. For each rational $r \in (0, 1)$, let F_r denote the family of all Solovay degrees $\mathbf{a}$ such that $\dim \mathbf{a} > r$. Then, for each rational $r \in (0, 1)$, F_r is a filter (Miyabe, Nies and Stephan [12] Theorem 5.1).

An analogous question is whether the Solovay degrees of qS-complete reals form a filter. The answer is affirmative.

Corollary 3.1. *(to Theorem 2.3) We consider the Solovay degrees of left-c.e. reals. Then the family of Solovay degrees of all qS-complete left-c.e. reals is a filter.*

Proof. The first requirement of a filter is obviously satisfied. As the degree c in the second requirement, by Theorem 2.3, we can take the Solovay degree of some Ω_T, where $T = 2^{-n}$ for a natural number n. $\qquad \square$

We are going to investigate ideals. By Theorem 1.1, the Solovay degrees of non-ML-random left-c.e. reals form an ideal. Miyabe, Nies and Stephan [12] Proposition 5.7 showed the following: For each $r \in [0, 1]$, the family of left-c.e. degrees $\mathbf{a}$ such that $\mathrm{Dim}(\mathbf{a}) < r$ is an ideal of left-c.e. Solovay degrees. The same thing holds for $\mathrm{Dim}(\mathbf{a}) \leq r$ in place of $\mathrm{Dim}(\mathbf{a}) < r$.

We ask whether the family of left-c.e. degrees of non-qS-complete left-c.e. degrees forms an ideal. The answer is negative.

Corollary 3.2. *(to Theorem 2.3) There are non-qS-complete left-c.e. reals α and β such that $\alpha + \beta$ (real addition) is qS-complete.*

Proof. The following was shown in [12] Theorem 4.1. Suppose that g is a computable function such that $\sum_n 2^{-g(n)}$ is finite and a computable real. For a string σ, let $C(\sigma)$ denote its plain Kolmogorov complexity (see Section 3.1 of Ref. 6). Let α be a left-c.e. real such that $C(\alpha \restriction n) \leq n - g(n)$ for all n. There exist left-c.e. reals β, γ such that $\alpha = \beta + \gamma$, $\dim(\beta) = \dim(\gamma) = 0$ (and both β, γ satisfy certain requirements).

For example, we look at the case of $\alpha = \Omega_{1/2}$ and $g(n) = n/2 - O(1)$. By Theorem 2.3, α is qS-complete, and neither β nor γ is qS-complete. □

3.2. *Rational sequences and reducibility*

By Corollary 3.2, the Solovay degrees of non-qS-complete left-c.e. reals do not form an ideal: Neither do the qS-degrees of them. Now we ask whether there is a stronger version $\ll_{qS}$ of qS-reducibility such that $\alpha, \beta \ll_{qS} \gamma \implies \alpha + \beta \ll_{qS} \gamma$. In this subsection, we investigate the relationships between computable rational sequences and reducibility. Based on these observations, we will see an example of a stronger version of qS-reducibility with the above-mentioned property in the next subsection. Solovay reducibility has many equivalent assertions. Downey *et al.* characterized it via rational sequences.

Lemma 3.1. *(Downey et al.[7]) Suppose $\langle b_n \rangle \nearrow \beta$. The following are equivalent.*

(1) $\alpha \leq_S \beta$
(2) $\exists \langle a_n \rangle \nearrow \alpha \quad \exists d > 0$ such that $\forall n \in \mathbb{N} \; a_n - a_{n-1} < d(b_n - b_{n-1})$.

In the case of quasi Solovay reducibility, the following holds.

Proposition 3.1. *The following are equivalent.*

(1) $\alpha \leq_{qS} \beta$

(2) $\exists \langle a_n \rangle \nearrow \alpha, \langle b_n \rangle \nearrow \beta \quad \exists d, \ell > 0$ *such that*
$\forall n, m \in \mathbb{N} \ (n < m \implies (a_m - a_n)^\ell < d(b_m - b_n))$.

Proof. The direction of (2) $\implies$ (1) is given by taking limit of $m \to \infty$. By carefully examining our construction of Hölder continuous function in Theorem 2 of (the preprint version of) Ref. 8, we show the direction of (1) $\implies$ (2). $\qquad \square$

In the case where β is 1-random, the rational sequences have more interesting properties. We are interested in the limit of $(\alpha - a_n)^\ell / (\beta - b_n)$ under the assumption that β is qS-complete.

Lemma 3.2. *For left-c.e. α and β, the following are equivalent.*

(1) $\alpha \leq_{qS} \beta$
(2) $\forall \langle b_s \rangle \nearrow \beta \quad \exists \langle a_s \rangle \nearrow \alpha \quad \exists d, \ell \in \mathbb{N}^+ \forall s \ (\alpha - a_s)^\ell \leq d(\beta - b_s)$
(3) $\forall \langle a_s \rangle \nearrow \alpha \quad \exists \langle b_s \rangle \nearrow \beta \quad \exists d, \ell \in \mathbb{N}^+ \forall s \ (\alpha - a_s)^\ell \leq d(\beta - b_s)$
(4) $\exists \langle b_s \rangle \nearrow \beta \quad \exists \langle a_s \rangle \nearrow \alpha \quad \exists d, \ell \in \mathbb{N}^+ \forall s \ (\alpha - a_s)^\ell \leq d(\beta - b_s)$

Proof. (1) $\implies$ (2): Let $a_s = f(b_s)$.
(1) $\implies$ (3): Given $\langle a_s \rangle$, take a temporary sequence $\langle \beta_s^* \rangle \nearrow \beta$. Let $N \in \mathbb{N}^+$ be large enough. For each $s \geq N$, take the largest t such that $f(\beta_t^*) \leq a_s$, and we define b_s as β_t^*. Then $(\alpha - a_s)^\ell \leq (\alpha - f(\beta_t^*))^\ell \leq d(\beta - b_s)$.
(2) $\implies$ (4), (3) $\implies$ (4): These are obvious.
(4) $\implies$ (1): Given $q < \beta$, find s such that $q \leq b_s$, and let $f(q) = a_s$. $\qquad \square$

The $\exists - \forall$ version of (2) is not equivalent to (1), because when both of α and β are rationals, we can chose arbitrarily slow $\{a_s\}$ afterword. Therefore, $\forall - \forall$ version of (2) is not equivalent to (1) in general. The situation is different with the hypothesis that β is qS-complete. We are going to see this in Lemma 3.3. The $\exists - \forall$ version of (3) is not equivalent to (1), because when both of α and β are rationals, we can chose arbitrarily fast $\{b_s\}$ afterword.

Lemma 3.3. *Suppose $\langle a_s \rangle \nearrow \alpha, \langle b_s \rangle \nearrow \beta$. If β is qS-complete, then there exist positive integers d, ℓ such that:*

$$\forall k \ (\alpha - a_k)^\ell \leq d(\beta - b_k)$$

Proof. Suppose that β is qS-complete. Then the statement of Lemma 3.2 (2) holds with $\alpha = \Omega$. That is, it holds that: $\exists \{\omega_s\} \nearrow \Omega$ $\exists d, \ell \in \mathbb{N}^+ \forall s \, (\Omega - \omega_s)^\ell \le d(\beta - b_s)$

On the other hand, by Theorem 1.2, $\dfrac{\alpha - a_s}{\Omega - \omega_s}$ has a limit. Therefore, there exists a positive integer e of the following property: $\forall s \; \alpha - a_s \le e(\Omega - \omega_s)$. Hence, it holds that $(\alpha - a_s)^\ell \le e^\ell(\Omega - \omega_s)^\ell \le e^\ell d(\beta - b_s)$. $\qquad\square$

Lemma 3.4. *Suppose* $\langle a_s \rangle \nearrow \alpha, \langle b_s \rangle \nearrow \beta$. *Suppose that* $x \ge 1$ *is a real number, and the following limit* > 0 *exists.*

$$\lim_{s \to \infty} \frac{(\alpha - a_s)^x}{\beta - b_s}$$

Then x *is uniquely determined (depending on* $\langle a_s \rangle$ *and* $\langle b_s \rangle$*).*

Proof. Suppose that the limit > 0 exists for x and that the same thing holds for $y > x$ in place of x, too. Then we have:

$$\frac{(\alpha - a_s)^y}{\beta - b_s} = (\alpha - a_s)^{y-x} \frac{(\alpha - a_s)^x}{\beta - b_s} \to 0 \; (s \to \infty)$$

This contradicts the assumption. $\qquad\square$

Theorem 3.1. *It holds that (1)* $\implies$ *(2)* $\implies$ *(3). If* β *is random, we also have (2)* $\impliedby$ *(3).*

   (1) $\alpha \ll_K \beta$. *To be more precise:* $\lim_{n \to \infty}(K(\beta \upharpoonright n) - K(\alpha \upharpoonright n)) = \infty$.

   (2) $\alpha \ll_S \beta$. *To be more precise:* $\forall \langle a_n \rangle \nearrow \alpha \quad \forall \langle b_n \rangle \nearrow \alpha$
$\lim_{n \to \infty} \dfrac{\alpha - a_n}{\beta - b_n} = 0$.
   (3) $\alpha <_S \beta$

We call $\ll_S$ above strong Solovay reducibility. This definition was clearly inspired by Theorem 1.2.

Proof. (1) $\implies$ (2): Assume (1). Given $\langle a_n \rangle$ and $\langle b_n \rangle$, apply Lemma 2.1 (1) to the case where $\ell = 1$. Then for almost all n, we have $(\alpha - a_n)/(\beta - b_n) \le 2/m$. Since m was arbitrary, we are done.
   (2) $\implies$ (3) is obvious.
   (3) and β is random $\implies$ (2): By Theorem 2.3 (b) of Ref. 11. $\qquad\square$

In particular, if $\beta = \Omega$ the assertions (2) and (3) of Lemma 3.1 are equivalent. In the proof of the following lemma, we will observe that given a left-c.e. real γ, the set of left-c.e. α such that $\alpha \ll_S \gamma$ forms an ideal. These two facts are important to see that the non-random left-c.e. reals form an ideal.

Lemma 3.5. *In Lemma 3.1, the following hold.*

(a) *If β is random then* (2) *implies the following* (1−).
  (1−) $\limsup_{n\to\infty} K(\beta \restriction n) - K(\alpha \restriction n) = \infty$
(b) *(2) does not imply (1).*
(c) *(3) does not imply (2).*

Proof. (a) Assume (2) of Lemma 3.1. By Ref. 2 (see also Ref. 11), α is not 1-random. Therefore, $\forall c \neg \forall^\infty n \; n - c \leq K(\alpha \restriction n)$. In other words, for all positive integer c, there are infinitely many n_c such that the following holds.

$$n_c - c > K(\alpha \restriction n_c) \tag{6}$$

Thus, there is an increasing sequence $\{n_c\}_{c \geq 1}$ of positive integers such that Eq. (6) holds for each c. Since β is random, for some positive integer d, it holds that $\forall n \; K(\beta \restriction n) > n - d$. Thus for each positive integer k, we have the following.

$$K(\beta \restriction n_k) - K(\alpha \restriction n_k) > (n_k - d) - (n_k - k) = k - d. \tag{7}$$

Hence, $\limsup_{n\to\infty} K(\beta \restriction n) - K(\alpha \restriction n) = \infty$.

(b) There exists a non-random left-c.e. real α such that $\liminf_n (K(\Omega \restriction n) - K(\alpha \restriction n)) < \infty$. Thus, $\alpha \not\ll_K \Omega$. On the other hand, by Ref. 2 (see the paragraph just after Theorem 1.2), we have $\alpha \ll_S \Omega$.

(c) Every non-random left-c.e. Solovay degree can split into lesser left-c.e. Solovay degrees (Downey *et al.*[7]. See also section 9.5 of Ref. 6). We are going to observe that this property of $<_S$ is not shared by $\ll_S$. Let α, β, and γ be left-c.e. reals such that $\alpha, \beta \ll_S \gamma$. Given $\langle a_s \rangle \nearrow \alpha$ and $\langle c_s \rangle \nearrow \gamma$, modify $\langle c_s \rangle$, if necessary, so that $\langle c_s - a_s \rangle$ is an increasing sequence of positive rationals. Let $b_s = c_s - a_s$. Then we have the following.

$$\frac{\alpha + \beta - (a_s + b_s)}{\gamma - c_s} = \frac{\alpha - a_s}{\gamma - c_s} + \frac{\beta - b_s}{\gamma - c_s} \to 0 \qquad (8)$$

Thus, it holds that $\alpha + \beta \ll_S \gamma$. We have shown $\alpha, \beta \ll_S \gamma \implies \alpha + \beta \ll_S \gamma$, therefore $<_S$ and $\ll_S$ are not equivalent for left-c.e. reals. However, we know that (2) implies (3). Hence (3) does not imply (2). $\qquad\square$

For any left-c.e. real α, the family of left-c.e. degrees $\leq_S \alpha$ forms an ideal, but the family of left-c.e. degrees $<_S \alpha$ does not form an ideal unless α is 1-random. By the proof of (c) above, we know that the family of left-c.e. degrees $\ll_S \alpha$ forms an ideal. The case $\alpha = \Omega$ corresponds to Theorem 1.1.

3.3. Ideals and a stronger version of qS-reducibility

Based on the observation in the previous subsection, we introduce a stronger version of qS-reducibility.

Definition 3.1. $\alpha \leqslant_{qS} \beta$ denotes the following assertion.

$$\exists \ell \in \mathbb{N} \; \forall \langle a_n \rangle \nearrow \alpha \; \forall \langle b_n \rangle \nearrow \beta \; \lim_{n \to \infty} \frac{(\alpha - a_n)^\ell}{\beta - b_n} = 0$$

The goal of this section is to show that $\alpha, \beta \leqslant_{qS} \gamma \implies \alpha + \beta \leqslant_{qS} \gamma$.

Theorem 3.2. *Suppose that β is left-c.e. Then we have (1q) $\implies$ (2q), and (2q) $\implies$ (3q). In addition, if β is qS-complete then we have (3q) $\implies$ (1q).*

(1q) *For some ℓ, $\lim_{n \to \infty} (K(\beta \restriction \ell n) - K(\alpha \restriction n)) = \infty$.*

(2q) $\alpha \leqslant_{qS} \beta$

(3q) $\alpha \leq_{qS} \beta$

Proof. (1q) $\implies$ (2q): Assume (1q). Apply Lemma 2.1 (1) to given $\langle a_n \rangle$ and $\langle b_n \rangle$. Then for almost all n, we have $(\alpha - a_n)^\ell / (\beta - b_n) \leq 2^\ell / m$. Since m was arbitrary, we are done.

(2q) $\implies$ (3q) is obvious.

(3q) and β is qS-complete $\implies$ (1q): We have $K(\alpha \restriction n) \leq K(\beta \restriction \ell n) + O(1)$. It is enough to show that for some $c > 0$ we have $\lim_{n \to \infty} (K(\beta \restriction cn) - K(\beta \restriction \ell n)) = \infty$. Let $d =$

$\liminf_{n\to\infty} K(\beta \upharpoonright n)/n$, and let $D = \limsup_{n\to\infty} K(\beta \upharpoonright n)/n$. Since β is qS-complete, by Theorem 2.3, d is positive. It is not hard to see that for any $\varepsilon > 0$, for sufficiently large n we have $d - \varepsilon < K(\beta \upharpoonright n)/n < 1 + \varepsilon$. Now, suppose $\varepsilon > 0$ is small enough. We take a sufficiently large integer k depending on ε. Then the following holds for all sufficiently large n.

$$\frac{K(\beta \upharpoonright k\ell n) - K(\beta \upharpoonright \ell n)}{\ell n} = k\frac{K(\beta \upharpoonright k\ell n)}{k\ell n} - \frac{K(\beta \upharpoonright \ell n)}{\ell n} \geq k(d - \varepsilon) - 1 - \varepsilon > 0 \tag{9}$$

Therefore it holds that $\lim_{n\to\infty}(K(\beta \upharpoonright k\ell n) - K(\beta \upharpoonright \ell n)) = \infty$. $\square$

Proposition 3.2. *In Theorem 3.2, we have the following.*

(a) *We can not replace* (3q) *by* $\alpha <_{qS} \beta$.
(b) (3q) *does not imply* (2q).
(c) (2q) *does not imply* (1q).

Proof. (a) Consider the case where $\alpha = \beta = \Omega$, $\ell = 2$, $\langle a_n \rangle \nearrow \Omega$ and $\langle b_n \rangle \nearrow \Omega$. By Ref. 2 (see also Ref. 11), $(\alpha - a_n)/(\beta - b_n) \to 0$ $(n \to \infty)$. Therefore, $(\alpha - a_n)^2/(\beta - b_n) \to 1$ $(n \to \infty)$. Thus, (2q) holds under the current assumptions. On the other hand, we have $\beta \leq_{qs} \alpha$, thus $\alpha <_{qS} \beta$ does not hold.

(b) Consider the case where both of α and β are computable. Then (3q) holds, and (2q) does not hold.

(c) follows from the two lemmas below. $\square$

Definition 3.2. A real α is a *strongly c.e. real* if there exists a c.e. set A of natural numbers and $\alpha = \sum_{n \in A} 2^{-n}$.

It is easy to see that any strongly c.e. real is left-c.e.

Lemma 3.6. *Let α be a computable real and β be a strongly c.e. real. Then, for all $\ell \in \mathbb{N}$ we have $K(\beta \upharpoonright \ell n) - K(\alpha \upharpoonright n) < O(1)$ for infinitely many n. Thus,* (1q) *does not hold.*

Proof. This fact immediately follows from the fact that any strongly c.e. real is infinitely often K-trivial by Proposition 2.2. in Ref. 3. For the sake of completeness, we give details here again.

Let $B \subseteq \mathbb{N}$ be an infinite c.e. set and β is the strongly c.e. real defined by B, that is, $\beta = \sum_{n \in B} 2^{-n}$. Take a computable sequence $\langle B_s \rangle$ of finite sets of natural numbers such that $B_s \subsetneq B_{s+1}$ and $B = \bigcup_s B_s$. Define $b_s = \sum_{n \in B_s} 2^{-n}$, which is the corresponding approximation of β. For each $n \in B$, There are infinitely many $n \in B$ such that n is enumerated into B at stage s and no $m \leq n$ is enumerated into B after the stage s, that is,

$$\exists^\infty n \exists s [n \in B_s \wedge \forall m < n (m \notin B_s \rightarrow \forall t > s \; m \notin B_s)] \qquad (10)$$

For such a pair n, s, we have $b_s \upharpoonright n = \beta \upharpoonright n$. Hence, $K(\beta \upharpoonright n) \leq^+ K(n)$ for infinitely many n. This implies $K(\beta \upharpoonright \ell n) \leq^+ K(\ell n) \leq^+ K(n)$ for infinitely many n.

On the other hand, since $K(n) \leq^+ K(\alpha \upharpoonright n)$, we have $\liminf_{n \to \infty} K(\beta \upharpoonright \ell n) - K(\alpha \upharpoonright n) < \infty$. Thus, (1q) does not hold for any computable real α and for any strongly c.e. real β. $\qquad \square$

Lemma 3.7. *There exists a strongly c.e. real β such that $\alpha \leqslant_{qS} \beta$ for every computable real. Thus, (2q) holds for this β and any computable real α.*

Our proof idea is as follows. We shall take a variant of the halting problem as a set B and let β be the corresponding real. Since B knows the strings with which the machine halts as inputs, one can compute a fast-growing function f that dominates all computable functions.

Our goal is to show

$$\lim_{s \to \infty} \frac{\alpha - a_s}{\beta - b_s} = 0. \qquad (11)$$

Otherwise, b_s is a good approximation of β for infinitely many s, and one can compute a sufficiently long initial segment of B, from which one can compute a function g such that $g(s) \geq f(s)$ for infinitely many s. This would contradict the property of f.

Proof. First, we construct the strongly c.e. real β. Let σ_n be the enumeration of $2^{<\omega}$ in the length-lexicographical order, that is, empty string, 0, 1, 00, 01, 10, 11, ... We define a c.e. set B by

$$B = \{n \in \mathbb{N} \; : \; U(\sigma_n) \downarrow\}$$

and let $\beta = \sum_{n \in B} 2^{-n}$.
We define a function $f : \mathbb{N} \to \mathbb{N}$ as follows.

$$f(n) := \max\{s \in \omega : \exists k \le 2n[U(\sigma_k) \text{ halts at stage } s.]\} \quad (12)$$

Here U is a fixed universal plain machine. By saying $U(\sigma)$ halts at stage s, we mean $U(\sigma)[s] \downarrow$ and $U(\sigma)[s-1] \uparrow$.

Claim 1: f dominates any computable function g, that is, $f(n) \ge g(n)$ for almost all n.

Fix an index e of a computable function such that Φ_e is total. Since U is universal, U can simulate $\Phi_e(n)$ within $C(\langle e, n \rangle) + O(1)$, where C denotes the plain Kolmogorov complexity. Since $C(\langle e, n \rangle) + O(1) = O(\log n)$, there exists a program σ_k with $k \le 2n$ such that $U(\sigma_k)$ simulate $\Phi_e(n)$ for all sufficiently large n. By the usual convention, the halting stage is larger than the output. Hence, f dominates Φ_e. This proves Claim 1.

We are going to show $(2q)$ $\alpha \ll_{qS} \beta$ with $\ell = 1$. Suppose $\langle a_s \rangle \nearrow \alpha$ and $\langle b_s \rangle \nearrow \beta$. Later we need to construct a *total* computable function g from $\langle b_s \rangle$. For infinitely many s, the term b_s is a good approximation, but for other s, the term b_s may not be a good approximation, which may prevent the totality of g. Thus, we first translate $\langle b_s \rangle$ into a well-behaved sequence $\langle d_n \rangle$.

Claim 2: From each computable sequence $\langle b_s \rangle$, one can compute a computable increasing sequence $\langle D_n \rangle$ of finite sets of natural numbers such that, letting $d_n = \sum_{m \in D_n} 2^{-m}$, we have $D_n \nearrow B$ and $\langle d_n \rangle \nearrow \beta$. Here, by $D_n \nearrow B$, we mean that D_n is increasing and $\lim_n D_n = B$.

Let $\langle B_s \rangle$ be an increasing computable sequence of finite sets of natural numbers converging to B. Let B'_s be the binary expansion of b_s, that is, $b_s = \sum_{n \in B'_s} 2^{-n}$. If there are two such sets, choose one of them as you wish. Given n, we can computably find $k \ge n$ and $s \ge n$ such that $B'_k \upharpoonright (2n+1) \subseteq B_s \upharpoonright (2n+1)$ because $B'_k \nearrow B$ and B is not computable. Then define $D_n := B'_k \upharpoonright (2n+1)$ for this k and define $\langle d_n \rangle$ as above accordingly. The resulting d_n may be smaller than b_k and b_n because we cut the B'_k to make D_n, but the difference is at most 2^{-2n-1}. Therefore we have $b_n \leq d_n + 2^{-2n-1}$ and

$$\beta - d_n \leq \beta - b_n + 2^{-2n-1} \tag{13}$$

Therefore, $\langle d_n \rangle \nearrow \beta$. This proves Claim 2.

Now we are ready to prove the lemma. For simplicity, we first observe the case where $\alpha - a_n < 2^{-2n}/n$. Assume for a contradiction that Eq. (11) fails:

$$\exists \varepsilon > 0 \; \exists^\infty n \; \frac{\alpha - a_n}{\beta - b_n} > \varepsilon \tag{14}$$

We define a function g by

$$g(n) := \max\{s \in \omega : \exists m \in D_n [U(\sigma_m) \text{ halts at stage } s.] \} + 1 \tag{15}$$

Since $D_n \subseteq B$, $U(\sigma_m)$ halts for every $m \in D_n$. Hence, g is a total computable function.

If Eq. (14) holds for some n, then we have $\beta - b_n < 1/(\varepsilon n 2^{2n})$ and $\beta - d_n < 2^{-2n}$ by Eq. (13). Then, $D_n \restriction 2n = B \restriction 2n$ and $g(n) = f(n) + 1$. Since there are infinitely many such n, this contradicts Claim 1.

In the general case, we need one more trick. Take a computable subsequence $\langle n_k \rangle$ such that $\alpha - a_{n_k} < 2^{-2k-2}/k$ for all k. We may have

$$\beta - d_n \leq \beta - b_n + 2^{-2k-1}$$

for $n_{k-1} < n \leq n_k$ by a similar construction in Claim 2. Now we define a total computable function g by

$$g(k) := \max_{n \in (n_{k-1}, n_k]} \max\{s \in \omega : \exists m \in D_n [U(\sigma_m) \text{ halts at stage } s.]\} + 1$$

If Eq. (14) holds for some n, then, by taking k such that $n_{k-1} < n \leq n_k$, we hae

$$\beta - d_n \leq \beta - b_n + 2^{-2k-1} \leq \beta - b_{n_{k-1}} + 2^{-2k-1}$$
$$< \frac{\alpha - a_{n_{k-1}}}{\varepsilon} + 2^{-2k-1} < \frac{1}{\varepsilon k 2^{2k}} + 2^{-2k-1} < 2^{-2k}.$$

For this n, k, we have $D_n \restriction 2k = B \restriction 2k$. Thus, there are infinitely many k such that $g(k) = f(k) + 1$, which contradicts Claim 1. $\qquad\square$

Lemma 3.8. $\alpha, \beta \ll_{qS} \gamma \Rightarrow \alpha + \beta \ll_{qS} \gamma$

Proof. Take appropreate $\langle a_n \rangle \nearrow \alpha, \langle b_n \rangle \nearrow \beta$ and ℓ. Then for any $\langle c_n \rangle \nearrow \gamma$, we have the following.

$$\lim_{n \to \infty} \frac{(\alpha - a_n)^\ell}{\gamma - c_n} = 0, \quad \lim_{n \to \infty} \frac{(\beta - b_n)^\ell}{\gamma - c_n} = 0$$

Here, we have the following.

$$((\alpha + \beta) - (a_n + b_n))^\ell = \sum_{k=0}^{\ell} \binom{\ell}{k} (\alpha - a_n)^k (\beta - b_n)^{\ell-k}$$

$$\leq \sum_{k=0}^{\ell} \binom{\ell}{k} (\max\{\alpha - a_n, \beta - b_n\})^\ell$$

$$\leq O(1)(\max\{\alpha - a_n, \beta - b_n\})^\ell$$

Therefore, $((\alpha + \beta) - (a_n + b_n))^\ell / (\gamma - c_n) \to 0 \ (n \to \infty)$. $\quad\square$

By these results above, we know that the family of left-c.e. degrees $\ll_{qS} \alpha$ forms an ideal. We also note that since $\leq_{qS}$ is a standard reducibility, the family of left-c.e. degrees $\leq_{qS} \alpha$ also forms an ideal.

4. Effective dimensions via the rate of convergence

By developing the previous section, we characterize the effective Hausdorff dimension of the left-c.e. reals by the rate of convergence of their computable approximations. In order to sketch the motive for the following theorem, let us observe the case where $(\alpha - a_s)^\ell / (\Omega - \omega_s)$ has a positive limit $c \leq 1$. For sufficiently large s, the approximate value of $\ell \log(\alpha - \alpha_s)$ is given by $\log(\Omega - \omega_s) + \log c$. Thus it is natural to look at the ratio $\log(\Omega - \omega_s) / \log(\alpha - \alpha_s)$ to investigate the role of ℓ in this context.

Theorem 4.1. *Let α be a left-c.e. real. Suppose that $\langle \omega_s \rangle \nearrow \Omega$ and $\langle a_s \rangle \nearrow \alpha$.*
(1)

$$\limsup_{s \to \infty} \frac{\log(\Omega - \omega_s)}{\log(\alpha - a_s)} = \mathrm{Dim}(\alpha) \quad (16)$$

(2)

$$\liminf_{s \to \infty} \frac{\log(\Omega - \omega_s)}{\log(\alpha - a_s)} = \dim(\alpha) \tag{17}$$

Remark: In the both assertions, the results are independent of the choice of $\langle \omega_s \rangle$ and $\langle a_s \rangle$.

Proof. (1) Let $d = \mathrm{Dim}(\alpha)$. For simplicity, we often omit the floor symbols $\lfloor \ \rfloor$. In the proof, a real number expressing the length of a string should be replaced by a certain integer.

Claim 1. $\limsup_s \log(\Omega - \omega_s)/\log(\alpha - a_s) \le d$.

Proof of Claim 1: Suppose $\varepsilon > 0$. We are going to show $\log(\Omega - \omega_s)/\log(\alpha - a_s) \le d + \varepsilon$ for sufficiently large s. For some constant c, for all n we have the following.

$$K(\Omega \upharpoonright (d + \varepsilon)n) > \lfloor (d + \varepsilon)n \rfloor - c \tag{18}$$

Given a positive integer N, all sufficiently large n satisfies the following.

$$\lfloor (d + \varepsilon)n \rfloor - c > \lfloor (d + \varepsilon/2)n \rfloor - N \tag{19}$$

Since $d = \mathrm{Dim}(\alpha)$, for almost all n, we have the following.

$$(d + \varepsilon/2)n > K(\alpha \upharpoonright n) \tag{20}$$

By Eq. (18), Eq. (19), and Eq. (20), for almost all n we have $K(\Omega \upharpoonright (d + \varepsilon)n) - K(\alpha \upharpoonright n) > N$. Since N was arbitarary, it holds that $\lim_{n \to \infty}(K(\Omega \upharpoonright (d + \varepsilon)n) - K(\alpha \upharpoonright n)) = \infty$.

Therefore, we can apply Lemma 2.1 (1) to this case with $\ell = d + \varepsilon$, and $m = \lfloor 2^{d+\varepsilon} \rfloor$. Let $\langle s_n \rangle$ be the sequence in Lemma 2.1 (1). Then for almost n and each s such that $s_n \le s \le s_{n+1}$, we have $\alpha - a_s \le 2^{-n}$ and $\Omega - \omega_s \ge 2^{-(d+\varepsilon)n}$. Noting that the divisor $\log(\alpha - a_s)$ is negative, we get $\limsup_s \log(\Omega - \omega_s)/\log(\alpha - a_s) \le d + \varepsilon$. Since $\varepsilon > 0$ was arbitrary, we have shown Claim 1. Q.E.D. (Claim 1)

Claim 2. $\limsup_s \log(\Omega - \omega_s)/\log(\alpha - a_s) \ge d$.

Proof of Claim 2: Suppose $\varepsilon > 0$. Given a positive integer N, all sufficiently large n satisfies the following.

$$K(\Omega \restriction (d - \varepsilon)n) + N < (d - \varepsilon/2)n \tag{21}$$

For infinitely many n, we have the following.

$$(d - \varepsilon/2)n < K(\alpha \restriction n) \tag{22}$$

By Eq. (21) and Eq. (22), for infinitely many n, it holds that $N < K(\alpha \restriction n) - K(\Omega \restriction (d - \varepsilon)n)$. Therefore, for infinitely many n, we have the following.

$$N < K(\alpha \restriction n/(d - \varepsilon)) - K(\Omega \restriction n) \tag{23}$$

Therefore we can apply Lemma 2.1 (2) to this case. The roles of α, β in the lemma are performed by Ω and α, respectively. Let $\ell = 1/(d - \varepsilon)$, and $m = 2^{1/(d-\varepsilon)}$. Let $\langle s_n \rangle$ be the sequence in Lemma 2.1 (2). There are infinitely many n such that for each s such that $s_n \leq s \leq s_{n+1}$, $\Omega - \omega_s \leq 2^{-n}$ and $\alpha - a_s \geq 2^{-n/(d-\varepsilon)}$. Thus, we have $\log(\Omega - \omega_s)/\log(\alpha - a_s) \geq d - \varepsilon$ infinitely often. Hence it holds that $\limsup_s \log(\Omega - \omega_s)/\log(\alpha - a_s) \geq d - \varepsilon$. Since $\varepsilon > 0$ was arbitrary, we have shown Claim 2. Q.E.D. (Claim 2)

By Claims 1 and 2, we have shown Eq. (16).

(2)   Let $d = \dim(\alpha)$.

Claim 3. $\liminf_s \log(\Omega - \omega_s)/\log(\alpha - a_s) \leq d$.

Proof of Claim 3: Suppose $\varepsilon > 0$. For some constant c, for all n we have Eq. (18). Given a positive integer N, all sufficiently large n satisfies Eq. (19). Infinitely many n satisfies Eq. (20). Therefore, for infinitely many n we have $K(\Omega \restriction (d + \varepsilon)n) - K(\alpha \restriction n) > N$. Since N was arbitrary, it holds that $\limsup_{n\to\infty}(K(\Omega \restriction (d + \varepsilon)n) - K(\alpha \restriction n)) = \infty$.

Therefore, we can apply Lemma 2.1 (2) to this case. We get $\liminf_s \log(\Omega - \omega_s)/\log(\alpha - a_s) \leq d + \varepsilon$. Since $\varepsilon > 0$ was arbitrary, we have shown Claim 3. Q.E.D. (Claim 3)

Claim 4. $\liminf_s \log(\Omega - \omega_s)/\log(\alpha - a_s) \geq d$.

Proof of Claim 4: Suppose $\varepsilon > 0$. Given a positive integer N, all sufficiently large n satisfies Eq. (21). For almost all n, we have

Eq. (22). Therefore for almost all n, it holds that $N < K(\alpha \upharpoonright n) - K(\Omega \upharpoonright (d - \varepsilon)n)$. Thus for almost all n, we have Eq. (23).

Therefore we can apply Lemma 2.1 (1) to this case. Letting $\langle s_n \rangle$ be the sequence in Lemma 2.1 (1), for almost all n and for each s such that $s_n \leq s \leq s_{n+1}$, $\Omega - \omega_s \leq 2^{-n}$ and $\alpha - a_s \geq 2^{-n/(d-\varepsilon)}$. Thus, for almost all n we have $\log(\Omega - \omega_s)/\log(\alpha - a_s) \geq d - \varepsilon$. Hence it holds that $\liminf_s \log(\Omega - \omega_s)/\log(\alpha - a_s) \geq d - \varepsilon$. Since $\varepsilon > 0$ was arbitrary, we have shown Claim 4. Q.E.D. (Claim 4)

By Claims 3 and 4, we have shown Eq. (17). □

5. Appendix: A variant of the first incompleteness theorem

Based on non-computability Ω, Chaitin[4] showed a variant of Gödel's first incompleteness theorem. Suppose that T is a consistent computable theory extending PA. Then there exists a natural number b such that for every natural number a, we have $T \nvdash b < K(a)$.

As a by-product of our observation, based on Solovay reducibility, we give a variant of Chaitin's incompleteness theorem. For simplicity, we use non-random left-c.e. real α and Ω. However, the proof works for any pair of left-c.e. reals α, β provided that β is not Solovay reducible to α. The absence of Solovay reduction plays the role of Gödel sentence. For each natural number m, we let $\overline{m}$ denote the numeral for m in a given formal system.

Proposition 5.1. *Suppose that T is a consistent c.e. theory extending PA. Suppose that β is a left-c.e. real that is not Solovay complete (among the left-c.e. reals). Assume that $\langle \omega_s \rangle, \langle q_s \rangle$ are computable (strictly) increasing sequences of rationals converging to Ω, and that $\langle b_s \rangle, \langle r_s \rangle$ are computable (strictly) increasing sequences of rationals converging to β. Then the following holds.*

For every positive integer L, there exists a natural number n such that for all natural numbers m, t,

$$T \nvdash \forall s \geq \bar{t} \left(|\omega_s - q_{\overline{m}}| < L|b_s - r_{\overline{n}}| \right) \tag{24}$$

Proof. We prove the proposition by contradiction. We assume that for some positive rational number L, for every natural number n

there exist natural numbers m, t with the following property.

$$T \vdash \forall s \geq \bar{t} \left(|\omega_s - q_{\overline{m}}| < L|b_s - r_{\overline{n}}| \right) \tag{25}$$

We fix L for a while. Let $\varphi(x, y, z)$ denote the following formula.

$$\forall s \geq x \left[|\omega_s - q_y| < L|b_s - r_z| \right] \tag{26}$$

Since T is c.e., we can effectively perform the following procedure: Given n, enumerate all proofs in T; if we find a proof of $\varphi(\bar{t}, \overline{m}, \overline{n})$ for some (t, m), output (t, m). By Eq. (25), each input n has an output.

To be more precise, there exists computable function $f : \mathbb{N} \implies \mathbb{N} \times \mathbb{N}$ such that for all natural number n we have the following. Here, $f : n \mapsto (t, m)$ and $1st((t, m)) = t$ and $2nd((t, m)) = m$.

$$T \vdash \forall s \geq \overline{1st(f(n))} \left(|\omega_s - q_{\overline{2nd(f(n))}}| < L|b_s - r_{\overline{n}}| \right) \tag{27}$$

Since T is Σ_1-sound, it is Π_1-sound, too. Therefore, the following holds (in the standsard model).

$$\forall s \geq 1st(f(n)) \left(|\omega_s - q_{2nd(f(n))}| < L|b_s - r_n| \right) \tag{28}$$

By taking the limit $s \implies \infty$, we get the following.

$$\Omega - q_{2nd(f(n))} \leq L(\beta - r_n). \tag{29}$$

Now, we replace L by a slightly larger rational number. We abuse notation and denote this new rational number by the same symbol L. Note that function $g(n) := 2nd(f(n))$ is computable. Then for all n we have $\Omega - q_{g(n)} < L(\beta - r_n)$. Therefore, it holds that $\Omega \leq_S \beta$.

This contradicts that Ω is Solovay complete, and β is not Solovay complete. $\square$

Proposition 5.2. *Let $\varphi(x, y, z)$ be the formula defined in the proof of Theorem 5.1.*

(1) *For any n there exists (t, m) such that $\mathbb{N} \models \varphi(\bar{t}, \overline{m}, \overline{n})$.*
(2) *T is incomplete.*

Proof. (1) We fix a positive rational L and take a natural number n as in the proof of Theorem 5.1. If we take m big enough, then we have $\Omega - q_m < L(\beta - r_n)$. Therefore $L(\beta - r_n) - (\Omega - q_m) > 0$. Let

$h(s)$ denote $L(b_s - r_n) - (\omega_s - q_m)$. Then we have $\lim_{s\to\infty} h(s) = L(\beta - r_n) - (\Omega - q_m) > 0$.

Hence there exists t such that the following holds.

$$\forall s \geq t \, \left(\omega_s - q_m < L(b_s - r_n)\right) \tag{30}$$

In summary, for any n there exist t and m such that we have $\mathbb{N} \models \varphi(\bar{t}, \overline{m}, \overline{n})$.

(2) Now, assume for a contradiction that $T \vdash \neg\varphi(\bar{t}, \overline{m}, \overline{n})$ holds for the above-mentioned t, m and n. Since $\varphi(\bar{t}, \overline{m}, \overline{n})$ is a Π_1-sentence, $\neg\varphi(\bar{t}, \overline{m}, \overline{n})$ is a Σ_1-sentence. Thus, by Σ_1-soundness of T, we have $\mathbb{N} \models \neg\varphi(\bar{t}, \overline{m}, \overline{n})$. This contradicts to (1). Hence it holds that $T \nvdash \neg\varphi(\bar{t}, \overline{m}, \overline{n})$. On the other hand, by Theorem 5.1, for any t, m, we have $T \nvdash \varphi(\bar{t}, \overline{m}, \overline{n})$. Hence T is incomplete. $\qquad\square$

Acknowledgments

The authors would like to thank participants of CTFM 2019 for their feedback. In particular, the authors are grateful to George Barmpalias for his insightful comments.

References

1. K.B. Athreya, J.M. Hitchcock, J.H. Lutz, and E. Mayordomo. Effective strong dimension in algorithmic information and computational complexity. *SIAM J. Comput.*, 37, (2007), 671–705. DOI: 10.1137/S0097539703446912

2. G. Barmpalias, and A. Lewis-Pye. Differences of halting probabilities. *J. Comput. Syst. Sci.*, 89, (2017), 349–360. DOI: 10.1016/j.jcss.2017.06.002

3. G. Barmpalias, and C.S. Vlek. Kolmogorov complexity of initial segments of sequences and arithmetical definability. *Theoret. Comput. Sci.*, 412, (2011), 5656–5667. DOI: 10.1016/j.tcs.2011.06.006

4. G.J. Chaitin. Information-theoretic limitations of formal systems. *Jour. ACM*, 21, (1974), 403–424.

5. O. Demuth. On constructive psudonumbers. Communicationes Mathematicae Universitatis Carolinae, 16, 1975, 315–331. https://hdl.handle.net/10338.dmlcz/105626

6. R.G. Downey, and D.R. Hirschfeldt. *Algorithmic Randomness and Complexity*. Springer, New York, 2010.

7. R.G. Downey, D.R. Hirschfeldt, and A. Nies. Randomness, computability, and density. *SIAM J. Comput.*, 31, (2002), 1169–1183.
   DOI: 10.1137/S0097539700376937
8. M. Kumabe, K. Miyabe, Y. Mizusawa, and T. Suzuki. Solovay reducibility and continuity. *Submitted to a journal.* Preprint, Solovay reduction and continuity. arXiv:1903.08625[math. L.O.] (2019).
9. A. Kučera, T. Slaman. Randomness and Recursive Enumerability *SIAM J. Comput.*, 31(1), (2001), 199–211.
   DOI: 10.1137/S0097539799357441
10. E. Mayordomo. A Kolmogorov complexity characterization of constructive Hausdorff dimension. *Inform. Process. Lett.*, 84, (2002), 1–3.
    DOI: 10.1016/S0020-0190(02)00343-5
11. J. Miller. On Work of Barmpalias and Lewis-Pye: A Derivation on the D.C.E. Reals. In: Computability and Complexity: Essays Dedicated to Rodney G. Downey on the Occasion of His 60th Birthday, 2017, 644–659.
    DOI: 10.1007/978-3-319-50062-1_39
12. K. Miyabe, A. Nies, and F. Stephan. Randomness and Solovay degrees. *Journal of Logic and Analysis*, 10:3, (2018), 1–13.
    DOI: 10.4115/jla.2018.10.3
13. A. Nies. *Computability and Randomness*. Oxford University Press, Oxford, 2009.
14. R.M. Solovay. Draft of paper (or series of papers) on Chaitin's work. Unpublished notes, May 1975, 215 pages.
15. K. Tadaki. A generalization of Chaitin's halting probability Ω and halting self-similar sets. *Hokkaido Math. J.*, vol. 31, (2002), 219–253.

Takeuti-Yasumoto Forcing Revisited

Satoru Kuroda*

*Department of Liberal Arts, Gunma Prefectural Women's University,
Tamamura, Gunma 370-1193, Japan
E-mail: satoru@mail.gpwu.ac.jp*

In this article, we develop the forcing method on nonstandard models of
bounded arithmetic which is initiated by G. Takeuti and M. Yasumoto.
We re-formalize their method in two-sort setting and construct generic
extensions for subclasses PTIME. We also relate separation problems of
complexity classes to properties of generic extensions.

Keywords: Bounded Arithmetic; Forcing.

1. Introduction

In late 1990's two consecutive papers by G. Takeuti and Y. Yasumoto[13,14] were published. These papers aimed at applying Boolean
valued model constructions for bounded arithmetic and relating them
to separation problems of complexity classes.

Prior to their work, forcing was used to construct models of
bounded arithmetic in different contexts. The first application of
forcing in bounded arithmetic was done by Paris and Wilkie,[11] who
proved that the theory IE_1 extended by a single predicate symbol
R does not prove that R is not a bijection from $n + 1$ to n. Their
construction was later extended to $I\Delta_0$ by M. Ajtai[1] and to Buss'
theory[3] by S. Riis.[12] This line of research is integrated into a more
general framework by Atserias and Müller.[2]

---
*This work was supported by JSPS KAKENHI Grant Number 18K03400. The
work was done partially while the author was visiting the Institute for Mathematical Sciences, National University of Singapore in 2017. The visit was supported
by the Institute.

Besides these results, J. Krajíček[9,10] gave different types of forcing construction for models of bounded arithmetic PV and V^1. Krajíček's motivation for these results is to provide nonstandard models of weak theories which satisfy unproven separation of complexity classes.

Then Takeuti and Yasumoto tried to give a comprehensive theory for forcing in nonstandard models of bounded arithmetic. Their main motivation was to relate separations of complexity classes in the standard world to generic models constructed from nonstandard models.

Although forcing type arguments seem very useful in bounded arithmetic just like in other branches of mathematical logic, there are few connections between the above mentioned results.

In this paper, we rearrange Takeuti-Yasumoto type forcing argument in a frame work which is different from the one they adopted. Namely we rework the construction of generic models in two sort language in Cook and Nguyen.[6]

In such a formulation we define generic extensions which correspond to subclasses of PTIME such as NC^1, L and NL. Moreover, we can establish connections between generic extensions and separation conditions of complexity classes as well as propositional proof systems in the ground model.

The main difficulty of our forcing argument lies in the fact that the forcing theorem holds only for Σ_0^B formulas. So we must compromise with rather ad hoc arguments to treat Σ_2^B or Π_2^B formulas. Still, we show that there are some connections between separation problems in the ground model and the behavior of Σ_2^B and Π_2^B formulas in the generic extensions.

Our results suggest that Takeuti-Yasumoto forcing works as a general framework for forcing constructions in bounded arithmetic.

The organization of the paper is as follows. In Section 2 we briefly review basic notions and results of bounded arithmetic which we will use in the later sections. In Section 3, we reformalize Takeuti-Yasumoto type forcing in the two-sort language and reprove some of their results. From Section 4 to Section 6, we define Boolean algebras and generic extensions for complexity classes NC^1, L and NL. In Section 7, we prove connections between generic extensions of subP classes and separation problems. Finally in Section 8, we argue the problem of how Σ_2^B and Π_2^B formulas behave in the generic extensions.

2. Preliminaries

In this paper we mainly treat two-sort bounded arithmetic theories which capture PTIME and its subclasses.

First, we briefly review basic notions of two-sort bounded arithmetic developed by Cook and Nguyen.[6] The two-sort language $\mathcal{L}_2^A$ comprises two sorts of variable: number variables denoted by lower case letters and string variables denoted by upper case letters. $\mathcal{L}_2^A$ also contains constant symbol 0, function symbols $S(x)$ (successor), $x + y$, $x \cdot y$, $|X|$ (the length of X) and relation symbols $x \leq y$ and $x \in Y$ (the xth bit of Y is 1). The relation $x \in Y$ is sometimes denoted as $Y(x)$. We will occasionally use the notation $X \in 2^a$ to denote that X is a string of length a.

Quantifiers of the form $\forall x \leq t$ and $\exists x \leq t$ are called bounded number quantifiers. Quantifiers of the form $\forall X \leq t$ and $\exists X \leq t$ are called bounded string quantifiers whose intended meanings are

$$\forall X \leq t \, \varphi(X) \Leftrightarrow \forall X(|X| \leq t \to \varphi) \text{ and } \exists X \leq t \, \varphi(X) \Leftrightarrow \exists X(|X| \leq t \wedge \varphi).$$

Σ_0^B is the set of $\mathcal{L}_2^A$ formulas in which all quantifiers are bounded number quantifiers. Σ_i^B is the set of $\mathcal{L}_2^A$ formulas which is defined by counting the alternations of bounded string quantifiers while ignoring bounded number quantifiers with a bounded existential string quantifier at the top. Π_i^B is defined similarly as Σ_i^B with a bounded universal string quantifier at the top.

An $\mathcal{L}_2^A$ structure is of the form (M_0, M) where M_0 and M denote the number part and the string part respectively.

$BASIC_2$ denotes a finite set of axioms which define symbols in $\mathcal{L}_2^A$. The base theory $\mathbf{V}^0$ is axiomatized by $BASIC_2$ together with

$$\Sigma_0^B\text{-COMP} : \forall a \exists X < a \forall y < a \, (\varphi(y) \leftrightarrow X(y)), \varphi(y) \in \Sigma_0^B.$$

$\mathbf{V}^i$ is the theory $\mathbf{V}^0$ extended by Σ_i^B-COMP.

In $\mathbf{V}^0$ we can code sequences of numbers as well as strings. In particular, there exists a Σ_0^B formula $Seq(X, n)$ which says that X is a sequence of n strings. If $Seq(X, n)$ then we denote the k-th string of X by $(X)_k$.

Cook and Nguyen established a general method for constructing two-sort theories for subclasses of PTIME by adding a single axiom to $\mathbf{V}^0$. Such axioms formalize complete problems for the complexity classes in concern.

For instance, a minimal theory for PTIME is defined by adding to $\mathbf{V}^0$ a single axiom which denotes that any monotone circuit can be evaluated. Specifically let

$$\delta_{MCV}(a, G, E, Y) \equiv \neg Y(0) \wedge Y(1) \wedge \forall x < a x \geq 2 \rightarrow$$
$$Y(x) \leftrightarrow \begin{bmatrix} (G(x) \wedge \forall y < x\ E(y, x) \rightarrow Y(y)) \vee \\ (\neg G(x) \wedge \exists y < x\ (E(y, x) \wedge Y(y))) \end{bmatrix}.$$

Definition 2.1. Define

$$MCV \equiv \forall a\ \forall G, E\ \exists Y < a + 2\ \delta_{MCV}(a, G, E, Y).$$

The $\mathcal{L}_2^A$ theory **VP** is axiomatized by $\mathbf{V}^0$ and MCV.

Theorem 2.1 (Cook-Nguyen[6]). *A function is Σ_1^B-definable in* **VP** *if and only if it is Σ_1^B-definable in* $\mathbf{V}^1$ *if and only if it is computable in polynomial time.*

NC^1 is the class of predicates which are decidable by uniform families of $O(\log n)$ depth Boolean circuits, or equivalently uniform families of Boolean formulas. For NC^1 the following formula defines the monotone formula value problem

$$\delta_{\mathrm{MFVP}}(a, G, I, Y) \equiv \forall x < a\ Y(x + a) \leftrightarrow I(x) \wedge 0 < x \rightarrow$$
$$Y(x) \leftrightarrow [(G(x) \wedge Y(2x) \wedge Y(2x + 1)) \vee (\neg G(x) \wedge (Y(2x) \vee Y(2x + 1)))].$$

Definition 2.2. Define

$$\mathrm{MFVP} \equiv \forall a\ \forall G, I\ \exists Y\ d\delta_{\mathrm{MFVP}}(a, G, I, Y).$$

The $\mathcal{L}_A^2$ theory $\mathbf{VNC}^1$ is axiomatized by $\mathbf{V}^0$ and MFVP.

L and NL are the class of predicates which are decidable by $O(\log n)$ space bounded deterministic and nondeterministic Turing machines respectively.

The theory for L is defined by axiomatizing the deterministic reachability problem. Specifically let

$$\delta_{\text{PATH}}(a, E, P) \equiv (P)^0 = 0 \land \forall v < a\; E((P)^v, (P)^{v+1}) \land (P)^{v+1} < a.$$

Then we define

Definition 2.3. Let PATH be the axiom

$$\text{Unique}(a, E) \to \exists P \leq \langle a, a \rangle\; \delta_{PATH}(a, E, P)$$

where

$$\text{Unique}(a, E) \equiv a \neq 0 \land \forall x < a\; \exists! y < a\; E(x, y).$$

VL is the $\mathcal{L}_A^2$ theory whose axioms are $\mathbf{V}^0$ and PATH.

For NL, we formalize reachability problems as follows:

$$\delta_{\text{CONN}}(a, E, Y) \equiv Y(0,0) \land \forall x < a\; (x \neq 0 \to \neg Y(0, x)) \land$$
$$\forall z < a\; \forall x < a\; Y(z+1, x) \leftrightarrow (Y(z, x) \lor \exists y < a Y(z, y) \lor E(y, x)).$$

Definition 2.4. Define

$$\text{CONN} \equiv \forall a\; \forall E\; \exists Y\; \delta_{MFVP}(a, E, Y).$$

The $\mathcal{L}_A^2$ theory **VNL** is axiomatized by $\mathbf{V}^0$ and CONN.

Theorem 2.2 (Cook-Nguyen[6]). *A function with polynomial growth is Σ_1^B-definable in* **VNC**^{1} *if and only if it is bitwise computable in* NC^1.

Theorem 2.3 (Cook-Nguyen[6]). *A function with polynomial growth is Σ_1^B-definable in* **VL** *if and only if it is bitwise computable in L.*

Theorem 2.4 (Cook-Nguyen[6]). *A function with polynomial growth is Σ_1^B-definable in* **VNL** *if and only if it is bitwise computable in NL.*

Complexity classes and bounded arithmetic theories are associated with propositional proof systems. We will review definitions and basic properties of some of such systems which will be used in the later section.

Definition 2.5. PK is the propositional fragment of LK. CPK is PK which operates on Boolean circuits instead of formulas.

Bounded string quantifiers are translated into quantifiers over propositional variables and bounded arithmetic theories are also associated with quantified propositional proof systems.

Σ_0^q is the set of propositional formulas. For $i \geq 1$, Σ_i^q and Π_i^q are defined by counting the alternations of quantifiers.

Definition 2.6. The quantified propositional proof system G is formalized by sequent calculus style. For $i \geq 0$, G_i is G with cut formulas restricted to $\Sigma_i^q \cup \Pi_i^q$ formulas. G_1^* is G_1 with tree-like proofs.

Definition 2.7. A propositional formula is Krom with respect to variables $x_1, \ldots, x_l$ if it is in conjunctive normal form and each x_i appears at most two clauses. A Σ_1^q-Krom formula is of the form

$$\exists x_1 \cdots \exists x_l \phi(x_1, \ldots, x_l, \bar{a})$$

where $\phi(x_1, \ldots, x_l, \bar{a})$ is Krom with respect to $x_1, \ldots, x_l$. G_{NL}^* is a subsystem of G_1^* where cut formulas are restricted to Σ_1^q-Krom.

For a bounded $\mathcal{L}_A^2$ formula $\varphi(\bar{x}, \bar{X})$, let $\|\varphi(\bar{x}, \bar{X})\|_{\bar{m},\bar{n}}$ denote its propositional translation as defined in Cook-Nguyen.[6]

Theorem 2.5 (Propositional translation Theorem I). *Let* $\varphi(\bar{x}, \bar{X}) \in \Sigma_0^B$. *If* $\mathbf{V}^1 \vdash \forall x \, \forall X \, \varphi(\bar{x}, \bar{X})$ *then* $\|\varphi(\bar{x}, \bar{X})\|_{\bar{m},\bar{n}}$ *have polynomial size* CPK *proofs for all* $\bar{m}, \bar{n} \in \omega$. *If* $\mathbf{VNC}^1 \vdash \forall x \, \forall X \, \varphi(\bar{x}, \bar{X})$ *then* $\|\varphi(\bar{x}, \bar{X})\|_{\bar{m},\bar{n}}$ *have polynomial size* PK *proofs for all* $\bar{m}, \bar{n} \in \omega$.

Note that for a propositional proof system F, there is a Σ_0^B formula $\mathrm{Prf}_{\mathcal{F}}(P, A)$ which represents that P is a $\mathcal{F}$-proof of the formula A.

Theorem 2.6. $\mathbf{V}^1$ *proves that* CPK *is sound.* $\mathbf{VNC}^1$ *proves that* PK *is sound.*

Theorem 2.7 (Propositional translation Theorem II). *Let* $\varphi(\bar{x}, \bar{X}) \in \Sigma_0^B$.

(1) If $\mathbf{V}^1 \vdash \forall x \, \forall X \, \varphi(\bar{x}, \bar{X})$ *then* $\|\varphi(\bar{x}, \bar{X})\|_{\bar{m},\bar{n}}$ *have poly-size* G_1^* *proofs.*

(2) If $\mathbf{VNC}^1 \vdash \forall x \, \forall X \, \varphi(\bar{x}, \bar{X})$ *then* $\|\varphi(\bar{x}, \bar{X})\|_{\bar{m},\bar{n}}$ *have poly-size* G_0^* *proofs.*

(3) If **VNL** $\vdash \forall x \, \forall X \, \varphi(\bar{x}, \bar{X})$ then $\|\varphi(\bar{x}, \bar{X})\|_{\bar{m}, \bar{n}}$ have poly-size G_{NL}^* proofs.

Theorem 2.8. $\mathbf{V}^1$ proves that G_1^* is sound.

Definition 2.8. The prenex Σ_1^q-witnessing problem for a quantified propositional proof system $\mathcal{F}$ is the following problem: given a Σ_1^q formula $\exists x_1 \cdots \exists x_k \phi(\bar{p}, x_1, \ldots, x_k)$, its F-proof P, and assignments $\bar{a}$ to $\bar{p}$, compute witnesses $b_1, \ldots, b_k$ such that $\phi(\bar{a}, b_1, \ldots, b_k)$ is true.

We will use the following fact in the later section.

Lemma 2.1. $\mathbf{V}^1$ proves that the prenex Σ_1^q-witnessing problem for G_0 is in FNC^1.

Proof. We will show that the following formalization of the statement is provable in $\mathbf{V}^1$. There exists a PV function $F(P, \phi, \bar{a})$ such that if P is a G_0-proof of $\phi = \exists x_1 \cdots x_k \phi_0(\bar{p}, x_1, \ldots, x_k)$ where ϕ_0 is a quantifier free formula and $\bar{a}$ are truth assignments to $\bar{p}$ then

$$\phi_0(\bar{p}, F(P, \phi, \bar{a})(0), \ldots, F(P, \phi, \bar{a})(k-1))$$

is true. We will formalize the proof in Cook and Morioka[5] by checking that each construction can be done by NC^1 functions and their validity can be proven in $\mathbf{V}^1$.

1) P-prototypes and Herbrand P-disjunctions.
   Let P be a G_0-proof of

   $$\longrightarrow \exists x_1 \cdots \exists x_k \phi(\bar{p}, x_1, \ldots, x_k)$$

   with variables as indicated. We may assume that P is in free variable normal form so that all free variables occurring in P are among $\bar{p}$. A P-prototype is a formula of the form $A(\bar{p}, B_1, \ldots, B_k)$ which is an auxiliary formula of some occurrence of $\exists$ introduction in P. Then it is easy to check that

   (1) Computing the list of all P-prototypes can be done by a TC^0-function, and
   (2) $\mathbf{V}^1$ prove that the sequent

   $$\longrightarrow \phi(\bar{p}, B_1^1, \ldots, B_k^1), \ldots, \phi(\bar{p}, B_1^m, \ldots, B_k^m)) \quad (*)$$

   is valid where $A_1, \ldots, A_m$ is the list of all P-prototypes.

We call the sequent ($*$) Herbrand P-disjunction. Note that the second fact follows from the observation that there is a TC^0 function $F(P)$ which for a G_0-proof of some prenex Σ_0^q-formula computes a PK proof of Herbrand P-disjunction provably in $\mathbf{V}^1$.

2) Extracting witnesses from Herbrand P-disjunction. For Herbrand P-disjunction ($*$) and $1 \leq j \leq k$ we define

$$E_j = \neg A_1 \wedge \cdots \wedge \neg A_{j-1} \wedge A_j.$$

Also for $1 \leq i \leq k$, define

$$\eta_i = \bigvee_{1 \leq i \leq m} (E_j \wedge B_i^j).$$

It is straightforward to check that these formulas can be constructed by TC^0 function. Also it is easily seen that for any assignment $\bar{a}$ there exists an unique $1 \leq j \leq m$ such that $\bar{a} \models E_j$ provably in $\mathbf{V}^1$.

3) Witnessing Σ_1^q theorems and proving the correctness.

Now we define the NC^1 function which extract the witness from a G_0-proof. Define $F(P, \phi, \bar{a})$ by

$$F(P, \phi, \bar{a}) = Y \Leftrightarrow |Y| = k \wedge \forall i < k(Y(i) \leftrightarrow \bar{a} \models \eta_i).$$

Since the Boolean sentence value problem is in NC^1 provably in $\mathbf{V}^1$, it follows that $F(P, \phi, \bar{a})$ is in FNC^1 and $\mathbf{V}^1$ proves this fact. $\qquad\square$

3. Takeuti-Yasumoto forcing in the two-sort language

In this section we redefine basic notions of generic constructions due to Takeuti and Yasumoto.[13]

Let $\mathfrak{M} = (M_0, M)$ be a structure of some two-sort language of bounded arithmetic. Throughout the paper, we assume that $\mathfrak{M}$ is a countable nonstandard model such that $\mathfrak{M} \models \neg Exp$ where $Exp \equiv \forall x \exists y \, (y = 2x)$ unless otherwise stated.

Example. The original Takeuti-Yasumoto forcing starts with the ground model which is defined in the following manner. Let $M \models Th(\mathbb{N})$ be a countable nonstandard model and fix $n \in M \setminus \omega$. Define

$$M^* = \{x \in M : x \leq \underbrace{n \# \cdots \# n}_{k \text{ times}} \text{ for some } k \in \omega\}$$

and

$$M_0 = \{|x| : x \in M^*\}.$$

We regard (M_0, M^*) as a two-sort structure by identifying each element $x \in M^*$ with its binary representation. It is easy to see that $(M_0, M^*) \models \mathbf{V}^\infty + \neg Exp$.

Let $n \in M_0 \setminus \omega$ and $\bar{p} = p_0, \ldots, p_{n-1}$ be the list of propositional variables coded by elements in $\mathfrak{M}$. We define C_n to be the set of Boolean formulas over variables from $\bar{p}$ coded in $\mathfrak{M}$. The precise definition of $\mathbb{C}_n$ can be found in Takeuti and Yasumoto.[13]

$\mathbb{C}_n$ can be regarded as a Boolean algebra with respect to either one of the following two partial orders:

$$C \leq_A C' \Leftrightarrow \forall X \, (|X| = n \to eval(C, X) \leq eval(C', X)),$$
$$C \leq_{CF} C' \Leftrightarrow \exists P \, \text{Prf}_{CF}(C \to C', P).$$

Where $eval(C, X)$ is a Σ_1^B predicate which expresses that X satisfies C. We also denote it by $X \models_A C$ or $X \models C$. Note that these two partial orders are identical only if Extended Frege is super. In the following definitions, the partial order $\leq$ on $\mathbb{C}_n$ represents either $\leq_A$ or $\leq_{CF}$.

Define $\mathbb{B}_C = \mathbb{C}_n / =_A$ and $\mathbb{B}_{CF} = \mathbb{C}_n / =_{CF}$. We omit the subscript and denote either Boolean algebra by $\mathbb{B}$ when there is no fear of confusion.

A sequence of elements of $\mathbb{B}$ in $\mathfrak{M}$ is referred to as $X : a \to \mathbb{B}$ where the ith element is denoted by $X[i]$. We define the set of such elements as

$$M^{\mathbb{B}} = \{X \in M : X : a \to \mathbb{B} \text{ for some } a \in M_0\}.$$

A set $\mathbb{I} \subseteq \mathbb{B}$ is an ideal if $0 \in \mathbb{I}$, $1 \notin \mathbb{I}$, it is closed under $\vee$ and downward closed with respect to the partial order.

An ideal $\mathbb{I}$ is M_0-complete if for any $X \in M^{\mathbb{B}}$ with $X : a \to \mathbb{B}$, if $X(y) \in \mathbb{I}$ for all $y < a$ then

$$\bigvee_{y < a} X(y) \in \mathbb{I}.$$

A set $\mathbb{G} \subseteq \mathbb{B}$ is a filter if $0 \notin \mathbb{G}$, $1 \in \mathbb{G}$, it is closed under $\wedge$ and upward closed with respect to the partial order. A filter $\mathbb{G} \subseteq \mathbb{B}$ is maximal if for any $X \in \mathbb{B}$, exactly one of X or $\neg X$ is in $\mathbb{G}$.

A set $\mathbb{D} \subseteq \mathbb{B}$ is dense over an M_0-complete ideal $\mathbb{I}$ if for any $X \in \mathbb{B} \setminus \mathbb{I}$ there is $X' \in \mathbb{D} \setminus \mathbb{I}$ such that $X' \leq X$. $\mathbb{D}$ is definable if there exists a formula φ such that

$$\mathbb{D} = \{X \in \mathbb{B} : \mathfrak{M} \models \varphi(X)\}.$$

A filter $\mathbb{G} \subseteq \mathbb{B}$ is a TY-generic over an M_0-complete ideal $\mathbb{I}$ if $(\mathbb{D} \setminus \mathbb{I}) \cap \mathbb{G} \neq \emptyset$ whenever $\mathbb{D}$ is dense over $\mathbb{I}$ and definable. We simply say that $\mathbb{G}$ is a TY-generic if it is a TY-generic maximal filter over some $\mathbb{I}$. Remark that a TY-generic $\mathbb{G}$ is not definable in $\mathfrak{M}$.

We define the generic model analogous to that in set theory.

For a TY-generic $\mathbb{G}$ over an M_0-complete ideal $\mathbb{I}$ and $X : a \to \mathbb{B}$ we define

$$i_{\mathbb{G}}(X) = \{y < a : X[y] \in \mathbb{G}\}.$$

We also denote $i_{\mathbb{G}}(X)$ by $X_{\mathbb{G}}$ elsewhere. Finally we define

$$M_{\mathbb{G}} = \{i_{\mathbb{G}}(X) : X \in M^{\mathbb{B}}\}.$$

Since the length of any $i_{\mathbb{G}}(X)$ is bounded by some element in M_0, we can regard the pair $(M_0, M_{\mathbb{G}})$ as a two-sort structure with a natural interpretation and denote it by $\mathfrak{M}[\mathbb{G}]$.

Theorem 3.1. *Let $\mathbb{I} \subseteq \mathbb{B}$ be an M_0-complete ideal and $X \in \mathbb{B} \setminus \mathbb{I}$. Then there exists a TY-generic $\mathbb{G} \subseteq \mathbb{B}$ such that $X \in \mathbb{G}$.*

Proof. Let $\mathbb{D}_0, \mathbb{D}_1, \ldots$ be an enumeration of all dense definable sets over $\mathbb{I}$ and $X_0 = X, X_1, \ldots$ be an enumeration of all elements of $\mathbb{B}$. We define $Y_0 = X, Y_1 \ldots \in \mathbb{B} \setminus \mathbb{I}$ inductively as follows.

(1) On odd stages, assume that $Y_0, Y_1, \ldots, Y_{2i}$ are already defined such that $Y_0 \geq Y_1 \geq \cdots \geq Y_{2i}$. Let $Y_{2i+1} \in \mathbb{D} \setminus \mathbb{I}$ be such that $Y_{2i+1} \leq Y_{2i}$.

(2) On even stages, assume that $Y_0, Y_1, \ldots, Y_{2i+1}$ are already defined such that $Y_0 \geq Y_1 \geq \cdots \geq Y_{2i}$. Then define

$$Y_{2i+2} = \begin{cases} Y_{2i+1} \wedge X_i, & \text{if } Y_{2i+1} \wedge X_i \notin \mathbb{I} \\ Y_{2i+1}, & \text{otherwise.} \end{cases}$$

Define

$$\mathbb{G} = \{Y \in \mathbb{B} : Y_i \leq Y \text{ for some } i \in \omega\}.$$

Then it is easy to see that $\mathbb{G}$ is a TY-generic over $\mathbb{I}$ and $X \in \mathbb{G}$. $\quad \square$

Note that the TY-generic $\mathbb{G}$ constructed in Theorem 3.1 satisfies $\mathbb{G} \cap \mathbb{I} = \emptyset$.

First we will show that $\mathfrak{M}[\mathbb{G}]$ is a model of the base theory $\mathbf{V}^0$.

Definition 3.1. Let $\varphi(\bar{x}, X_1, \ldots, X_k) \in \Sigma_0^B$ and

$$\|\varphi(\bar{x}, X_1, \ldots, X_k)\|_{\bar{m}, n_1, \ldots, n_k}(\bar{p}_1, \ldots, \bar{p}_k)$$

be its propositional translation defined in $\mathfrak{M}$ where $\bar{p}_i$ corresponds to the variable X_i. For $a \in M_0$ and $A_1, \ldots, A_k \in M^{\mathbb{B}}$ with $A_i : b_i \to \mathbb{B}$, we define

$$[\![\varphi(\bar{a}, \bar{A})]\!] = \|\varphi(\bar{x}, \bar{X})\|_{\bar{a}, b_1, \ldots, b_k}(X_1, \ldots, X_k)$$

In the following sections we will construct Boolean algebras for other complexity classes and prove forcing theorem for those algebras. The key point of proving them is that they admit similar translations of Σ_0^B formulas.

Theorem 3.2 (Forcing Theorem for Σ_0^B formulas). *Let* $\mathfrak{M} \models$ $\mathbf{V}^1 \; \varphi(\bar{x}, X_1, \ldots, X_k)$ *be a Σ_0^B formula with parameters as indicated. Then for any $\bar{a} \in M_0$ and $A_1, \ldots, A_k \in M^{\mathbb{B}}$*

$$\mathfrak{M}[\mathbb{G}] \models \varphi(\bar{a}, i_{\mathbb{G}}(A_1), \ldots, i_{\mathbb{G}}(A_k)) \Leftrightarrow [\![\varphi(\bar{a}, A_1, \ldots, A_k)]\!] \in \mathbb{G}$$

for any TY-generic $\mathbb{G}$ over an M_0-complete ideal $\mathbb{I}$.

The proof of Theorem 3.2 is given in Takeuti and Yasumoto.[13]

Theorem 3.3. *Let* $\mathfrak{M} \models \mathbf{V}^1$. *If $\mathbb{G} \subseteq \mathbb{B}$ is a TY-generic over an M_0-complete ideal $\mathbb{I} \subseteq \mathbb{B}$ then $\mathfrak{M}[\mathbb{G}] \models \mathbf{V}^0$.*

The proof of Theorem 3.3 can be found in Takeuti and Yasumoto.[13]

The following fact is also useful.

Theorem 3.4. *Let* $\mathbb{B}$ *be either* $\mathbb{B}_C$ *or* $\mathbb{B}_{CF}$ *and* $\mathbb{G} \subseteq \mathbb{B}$ *be a TY-generic. Then $\mathbb{G}$ is closed under CF-provability, that is if $X_1, \ldots, X_k \in \mathbb{G}$ with $k \in M_0$ and there is a CF-proof of X from $X_1, \ldots, X_k$ in M then $X \in \mathbb{G}$.*

Proof. Let $X_1, \ldots, X_k$ and X be as above. Then the claim for $\mathbb{B}_{CF}$ follows from

$$\bigwedge_{1 \leq i \leq k} X_i \leq_{CF} X$$

and

$$\bigwedge_{1 \leq i \leq k} X_i \in \mathbb{G}.$$

For $\mathbb{B}_C$, it suffices to replace $\leq_{CF}$ of the first condition by $\leq_C$ using the soundness of CF in $\mathfrak{M}$. □

Theorem 3.5. *Let* $\mathfrak{M} \models \mathbf{V}^1$ *and* $\varphi(\bar{x}, \bar{X}) \in \Sigma_0^B$. *If for any* $\bar{a}, \bar{m} \in M_0$

$$\mathfrak{M} \models \exists P \; \mathrm{Prf}_{CF}(P, \|\varphi(\bar{x}, \bar{X})\|_{\bar{a}, \bar{m}})$$

then

$$\mathfrak{M}[\mathbb{G}] \models \forall x \forall X \; \varphi(\bar{x}, \bar{X})$$

for any TY-generic $\mathbb{G} \subseteq \mathbb{B}$.

Proof. Let $\varphi(\bar{x}, \bar{X})$ be as above and $P \in M$ be a CF-proof of

$$\|\varphi(\bar{x}, X_1, \ldots, X_k)\|_{\bar{a}, \bar{m}}(\bar{x}_1, \ldots, \bar{x}_l).$$

Let $\bar{a} \in M_0$ and $A_1, \ldots, A_k \in M^{\mathbb{B}}$ with $A_i : m_i \to \mathbb{B}$ for $1 \leq i \leq k$. From P we can construct a CF-proof of

$$[\![\varphi(\bar{a}, A_1, \ldots, A_k)]\!] = \|\varphi(\bar{x}, X_1, \ldots, X_k)\|_{\bar{a}, \bar{m}}(\bar{A}_1, \ldots, \bar{A}_l)$$

in $\mathfrak{M}$. Thus we have

$$[\![\varphi(\bar{a}, A_1, \ldots, A_k)]\!] \in \mathbb{G}$$

for any TY-generic $\mathbb{G} \subseteq \mathbb{B}_{CF}$. If $\mathbb{G} \subseteq \mathbb{B}_C$ then the claim follows from the fact that $\mathbf{V}^1$ proves the reflection principle for CF. □

For the Boolean algebra of circuits, we can show that the generic extension is a model of a theory for PTIME.

Theorem 3.6. *Let* $\mathfrak{M} \models \mathbf{V}^1$ *and* $\mathbb{B}$ *be either* $\mathbb{B}_C$ *or* $\mathbb{B}_{CF}$. *If* $\mathbb{G} \subseteq \mathbb{B}$ *is a TY-generic then* $\mathfrak{M}[\mathbb{G}] \models \mathbf{VP}$.

Proof. It suffices to show that $\mathfrak{M}[\mathbb{G}] \models MCV$. Let $\mathbb{B} = \mathbb{B}_{CF}$, $a \in M_0$, $G : a \to \mathbb{B}$ and $E : \langle a, a \rangle \to \mathbb{B}$. Define $Y : a + 2 \to \mathbb{B}$ inductively as follows:

$$Y[0] = 0, Y[1] = 1$$

$$Y[x] = \left(G[x] \wedge \bigwedge_{y<x} (E[y,x] \to Y[y]) \right) \vee \left(\neg G[x] \wedge \bigvee_{y<x} (E[y,x] \wedge Y[y]) \right).$$

We claim that $[\![\delta_{MCV}(a,G,E,Y)]\!] =_{CF} 1$. Remark that

$$[\![\delta_{MCV}(a,G,E,Y)]\!]$$
$$= \neg Y[0] \wedge Y[1] \wedge$$
$$\bigwedge_{2 \leq x < a+2} \left\{ Y[x] \leftrightarrow \left(\begin{pmatrix} G[x] \wedge \bigwedge_{y<x} (E[y,x] \to Y[y]) \end{pmatrix} \vee \\ \begin{pmatrix} \neg G[x] \wedge \bigvee_{y<x} (E[y,x] \wedge Y[y]) \end{pmatrix} \right) \right\}.$$

Trivially there is a CF-proof of $\neg Y[0]$ and $Y[1]$. Moreover there is a PTIME procedure which produces CF-proofs of

$$Y[x] \leftrightarrow \left(\begin{pmatrix} G[x] \wedge \bigwedge_{y<x} (E[y,x] \to Y[y]) \end{pmatrix} \vee \\ \begin{pmatrix} \neg G[x] \wedge \bigvee_{y<x} (E[y,x] \wedge Y[y]) \end{pmatrix} \right)$$

for $x < a$ in $\mathfrak{M}$.

If $\mathbb{B} = \mathbb{B}_C$ then we can evaluate the circuit computing MCV defined as above in $\mathfrak{M}$. So we have $[\![\delta_{MCV}(a,G,E,Y)]\!] =_{CF} 1$. $\quad\square$

A set $\mathbb{S} \subseteq \mathbb{B}$ is consistent if there is $A \in 2^n$ such that $A \models X$ for all $X \in \mathbb{S}$.

For a propositional proof system $\mathcal{F}$, a set $\mathbb{S} \subseteq \mathbb{B}$ is $\mathcal{F}$-consistent if there is no $\mathcal{F}$-proof of $\bot$ from $\mathbb{S}$ in $\mathfrak{M}$.

Theorem 3.7. *Let $\mathbb{S} \subseteq \mathbb{B}_C$ be a consistent set in $\mathfrak{M}$. Then there exists a TY-generic $\mathbb{G} \subseteq \mathbb{B}_C$ such that $\mathbb{S} \subseteq \mathbb{G}$.*

Proof. Let $\mathbb{S} \subseteq \mathbb{B}$ be a consistent set and define

$$\mathbb{I}_{\mathbb{S}} = \{ X \in \mathbb{B} : \mathbb{S} \cup \{X\} \text{ is inconsistent } \}.$$

We show that $\mathbb{I}_{\mathbb{S}}$ is an M_0-complete ideal. Trivially, $0 \in \mathbb{I}_{\mathbb{S}}$ and $1 \notin \mathbb{I}_{\mathbb{S}}$. If $X \in \mathbb{I}_{\mathbb{S}}$ and $X' \leq X$ then $X' \in \mathbb{I}_{\mathbb{S}}$ since

$$\forall \alpha \in 2^n \ (\alpha \models X' \to \alpha \models X).$$

If $X, X' \in \mathbb{I}_\mathbb{S}$ then trivially $X \vee X' \in \mathbb{I}_\mathbb{S}$. The M_0-completeness is proved in a similar manner.

Now note that for each $X \in \mathbb{S}$, $\neg X \in \mathbb{I}_\mathbb{S}$. Let $\mathbb{G}$ be a TY-generic such that $\mathbb{G} \cap \mathbb{I}_\mathbb{S} = \emptyset$. Then by the maximality of $\mathbb{G}$ it must be that $X \in \mathbb{G}$ for all $X \in \mathbb{S}$. □

Theorem 3.8. *Let* $\mathbb{S} \subseteq \mathbb{B}_{CF}$ *be a CF-consistent set in* $\mathfrak{M}$. *Then there exists an TY-generic* $\mathbb{G} \subseteq \mathbb{B}_{CF}$ *such that* $\mathbb{S} \subseteq \mathbb{G}$.

Proof. Let $\mathbb{S} \subseteq \mathbb{B}_{CF}$ be CF-consistent and

$$\mathbb{I}_\mathbb{S} = \{X \in \mathbb{B} : \mathbb{S} \cup \{X\} \text{ is CF-inconsistent}\}.$$

We claim that $\mathbb{I}_\mathbb{S}$ is an M_0-complete ideal. Trivially, $0 \in \mathbb{I}_\mathbb{S}$ and $1 \notin \mathbb{I}_\mathbb{S}$. If $X \in \mathbb{I}_\mathbb{S}$ and $X' \leq X$ then there exist CF-proofs P of $\bot$ from $\mathbb{S}$ and P' of X from X' and from these proofs we can construct a CF-proof of $\bot$ from $\mathbb{S} \cup \{X'\}$. Thus we have $X' \in \mathbb{I}_\mathbb{S}$. If $X, X' \in \mathbb{I}_\mathbb{S}$ then there are CF-proofs P and P' of $\bot$ from $\mathbb{S} \cup \{X\}$ and $\mathbb{S} \cup \{X'\}$ respectively and from these proof we can construct a CF-proof of $\bot$ from $\mathbb{S} \cup \{X' \vee X\}$. The M_0-completeness is proved in a similar manner.

The rest of the proof proceeds as in the proof of Theorem 3.7. □

4. Boolean valued models for NC^1

In the following three sections we will define Boolean algebras which corresponds to subclasses of P. To begin with, we define Boolean algebras for NC^1.

The following arguments can be formalized in some model $\mathfrak{M}$ mentioned as above.

We construct a Boolean algebra which consists of Boolean formulas in $\mathfrak{M}$. Define

$Formula(G, E) \Leftrightarrow$
$E \subseteq |G| \times |G| \wedge \forall x, y, z < |G| \ (x < y \wedge x < z \wedge E(x, y) \wedge E(x, z)) \to y = z.$

and set

$$\mathbb{F} = \{\langle G, E \rangle \in M : \mathfrak{M} \models Formula\langle G, E \rangle\}.$$

For $(G, E) \in \mathbb{F}$ and $A \in 2^n$ we define the function $FEval(A, G, E)$ as

$$
\begin{aligned}
FEval(A, G, E) = Y \Leftrightarrow |Y| = |G| \wedge \forall x < |G| \\
Y(x) \leftrightarrow (x < n \wedge A(x)) \vee \\
(n \leq x \wedge ((G(x) \wedge \forall y < x\ (E(y, x) \rightarrow Y(y)))) \vee \\
(\neg G(x) \wedge \exists y < x\ (E(y, x) \wedge Y(y))))).
\end{aligned}
$$

Lemma 4.1. *The function $FEval$ is Σ_0^B-definable in $\mathbf{VNC}^1$.*

Proof Sketch. In $\mathbf{VNC}^1$, we can transform any Boolean formula into a balanced Boolean formula. So we can use MFVP to define $FEval$. □

Define

$$
A \models_F \langle G, E \rangle \Leftrightarrow |G| \in FEval(A, G, E).
$$

Let $\langle G, E \rangle, \langle G', E' \rangle \in \mathbb{F}$. We define the partial order $\leq_F$ as

$$
\langle G, E \rangle \leq_F \langle G', E' \rangle \Leftrightarrow \forall A \in 2^n (A \models_F \langle G, E \rangle, \rightarrow A \models_F \langle G', E' \rangle).
$$

and define

$$
\mathbb{B}_F = \mathbb{F} / =_F .
$$

Since NC^1 corresponds to the propositional proof system PK and G_0, we can define another Boolean algebra based on the propositional provability.

Let $\langle G, E \rangle, \langle G', E' \rangle \in \mathbb{F}$. We define the partial order $\leq_{G_0}$ as

$$
\langle G, E \rangle \leq_{G_0} \langle G', E' \rangle \Leftrightarrow \exists P\ \mathrm{Prf}_{G_0}(P, \langle G, E \rangle \rightarrow \langle G', E' \rangle).
$$

Finally, define

$$
\mathbb{B}_{G_0} = \mathbb{F} / =_{G_0} .
$$

Lemma 4.2. *Let $\mathfrak{M} \models \mathbf{VNC}^1$, $\varphi(\bar{x}, \bar{X}) \in \Sigma_0^B$ and $\bar{l}, \bar{m} \in M_0$. There exists $G, E \in M$ such that $\langle G, E \rangle$ is a formula in $\mathfrak{M}$ and*

$$
\mathfrak{M} \models \forall A_1, \ldots, A_k\ (\langle A_1, \ldots, A_k \rangle \models \|\varphi(\bar{x}, \bar{X})\|_{\bar{l}, \bar{m}} \leftrightarrow A \models_F \langle G, E \rangle).
$$

Proof. By structural induction on φ. □

So we identify the propositional translation $\|\varphi\|$ to the equivalent formula as given by Lemma 4.2.

Definition 4.1. Let $\mathbb{B}$ be either $\mathbb{B}_F$ or $\mathbb{B}_{G_0}$. For $\varphi(\bar{x}, X_1, \ldots, X_k) \in \Sigma_0^B$, $\bar{a} \in M_0$ and $A_1, \ldots, A_k \in M^{\mathbb{B}}$ with $A_i : b_i \to \mathbb{B}$ with $b_1, \ldots, b_k \in M_0$, we define

$$[\![\varphi(\bar{a}, A_1, \ldots, A_k)]\!] = \|\varphi(\bar{x}, X_1, \ldots, X_k)\|_{\bar{a}, b_1, \ldots, b_k}.$$

Theorem 4.1 (Forcing Theorem for NC^1). *Let* $\mathfrak{M} \models \mathbf{VNC}^1$, $\mathbb{B}$ *be either* $\mathbb{B}_F$ *or* $\mathbb{B}_{G_0}$, $\varphi(\bar{x}, X_1, \ldots, X_k) \in \Sigma_0^B$, $\bar{a} \in M_0$ *and* $A_1, \ldots, A_k \in M^{\mathbb{B}}$. *Then for any TY-generic* $\mathbb{G} \subseteq \mathbb{B}$

$$\mathfrak{M}[\mathbb{G}] \models \varphi(\bar{a}, i_{\mathbb{G}}(A_1), \ldots, i_{\mathbb{G}}(A_k)) \Leftrightarrow [\![\varphi(\bar{a}, A_1, \ldots, A_k)]\!] \in \mathbb{G}$$

The proof of Theorem 4.1 is identical to that for Theorem 3.2.

Theorem 4.2. *If* $\mathfrak{M} \models \mathbf{VP}$ *and* $\mathbb{G} \subseteq \mathbb{B}_F$ *is a TY-generic then* $\mathfrak{M}[\mathbb{G}] \models \mathbf{VNC}^1$.

Proof. We will show that

$$\mathfrak{M}[\mathbb{G}] \models \forall a \, \forall G \, \forall I \, \exists Y \, \delta_{\mathrm{MFVP}}(a, G, I, Y).$$

Recall that

$\delta_{\mathrm{MFVP}}(a, G, I, Y) \equiv$
$\qquad \forall x < a \, (Y(x + a) \leftrightarrow I(x)) \wedge$
$$0 < \forall x < a \left\{ Y(x) \leftrightarrow \begin{array}{l} ((G(x) \wedge Y(2x) \wedge Y(2x + 1)) \vee \\ (\neg G(x) \wedge (Y(2x) \vee Y(2x + 1)))) \end{array} \right\}.$$

Let x_i^I, x_i^G for $i < a$ be propositional variables and for $i < a$, define ϕ_i as

$$\phi_{i+a} = x_i^I,$$
$$\phi_i = (x_i^G \wedge \phi_{2i} \wedge \phi_{2i+1}) \vee (\neg x_i^G \wedge (\phi_{2i} \vee \phi_{2i+1})).$$

Let $a \in M_0$ and $G, I : a \to \mathbb{B}$ and define $Y : 2a \to \mathbb{B}$ as

$$Y[i] = \phi_i(G, I).$$

Then it is easy to show that

$$[\![\delta_{\mathrm{MFVP}}(a, G, I, Y)]\!] =_{G_0} 1.$$

and by the soundness of G_0 in $\mathfrak{M}$ we have

$$[\![\delta_{\mathrm{MFVP}}(a, G, I, Y)]\!] =_F 1. \qquad \square$$

5. Boolean valued models for L

We will use the fact that L is captured by polynomial size branching programs in order to construct the Boolean algebra for L.

A branching program with n inputs is a rooted directed acyclic graph such that

(1) each non-sink node is labelled by p_i for some $i < n$,
(2) each sink is labelled by either 0 or 1,
(3) each non-sink node has exactly two out-edges each labelled by 0 and 1.

We code branching programs by sequences as follows. Let $\bar{p} = p_0, \ldots, p_{n-1}$ be list of propositional variables. A branching program with l nodes and n inputs is a sequence $X = \langle v_0, \ldots, v_{l-1} \rangle$ where each v_i is either 0, 1 or of the form $\langle j, k_0, k_1 \rangle$ where $j < n$ and $k_0, k_1 > i$. It is easy to see that there exists a Σ_0^B formula $BP(n, X)$ which says that X is a branching program with n inputs.

Let X be such that $BP(n, X)$ and $A \in 2^n$. The computation of X on input A is given by the unique path $P = \langle a_0, \ldots, a_l \rangle$ such that

$$a_0 = 0 \wedge$$
$$\forall x < m((X_{a_x})_0 \in A \wedge a_{x+1} = (X_{a_x})_1) \vee ((X_{a_x})_0 \notin A \wedge a_{x+1} = (X_{a_x})_2)$$
$$\wedge (X_{a_m} = 0 \vee X_{a_m} = 1).$$

Let $CPATH(n, A, X, P)$ denote this formula. Then we have

Lemma 5.1. VL *proves that*

$$\forall n \forall A \in 2^n \forall X \, (BP(n, X) \to \exists P \, CPATH(n, A, X, P)).$$

Proof. We argue in **VL**. We define a Σ_0^B formula $\varphi(n, x, y, A, X)$ such that if $BP(n, X)$ and $A \in 2^n$ then φ defines a subgraph X' of X so that

- $Unique(Len(X), X')$ and
- the unique path from the root to a sink in X' is the accepting path of X on input A.

Specifically, let

$$\varphi(n, x, y, A, X) \equiv (A(((X)_x)_0) \wedge ((X)_x)_1 = y) \vee (\neg A(((X)_x)_0) \wedge ((X)_x)_2 = y).$$

By Σ_0^B-COMP, we have a sequence X' such that

$$\forall x, y \ (X'(x, y) \leftrightarrow \varphi(n, x, y, A, X))$$

and it is easy to see that $Unique(m, X')$ for some m. Thus PATH axiom yields the unique computation path of X on input A. □

Definition 5.1. For X with $BP(n, X)$ and $X \in 2^n$ we define

$$A \models_{BP} X \Leftrightarrow \exists P \ (CPATH(n, A, X, P)) \wedge (P)_{Len(P)-1} = 1).$$

Let $\mathfrak{M} \models \mathbf{VL}$ and $\bar{p} = p_0, \ldots, p_{n-1}$ be a list of propositional variables in $\mathfrak{M}$. We define

$$BP(\bar{p}) = \{X \in M : BP(n, X)\}.$$

For $X, X' \in BP(\bar{p})$, we define

$$X \leq_{BP} X' \Leftrightarrow \forall A \in 2^n \ (A \models_{BP} X \to A \models_{BP} X')$$

and define $\mathbb{B}_{BP} = BP(\bar{p})/ =_{BP}$.

In order to define the translation of Σ_0^B formulas into branching programs, we first define Boolean operations over branching programs. Let X be a branching program and k be a number. The sequence $X[+k]$ is obtained from X by incleasing all indices of nodes by k. Let X, Y be branching programs with $Len(X) = k$. Let us call nodes with labels 0 and 1 by 0-sink and 1-sink respectively. We define

(1) $X \wedge X'$ is obtained by replacing the 1-sink of X by the root of $Y[+k]$ and append $Y[+k]$ to X.
(2) $X \vee X'$ is obtained by replacing the 0-sink of X by the root of $Y[+k]$ and append $Y[+k]$ to X.
(3) $\neg X$ is obtained from X by swapping the 0-sink and the 1-sink.

It is easy to see that these operations are AC^0 computable.

Theorem 5.1 (Forcing Theorem for L). *Let* $\mathfrak{M} \models \mathbf{VL}$, $\varphi(\bar{x}, X_1, \ldots, X_k) \in \Sigma_0^B$, $\bar{a} \in M_0$ *and* $A_1, \ldots, A_k \in M^{\mathbb{B}}$. *Then for any TY-generic* $\mathbb{G} \subseteq \mathbb{B}_{BP}$

$$\mathfrak{M}[\mathbb{G}] \models \varphi(\bar{a}, i_{\mathbb{G}}(A_1), \ldots, i_{\mathbb{G}}(A_k)) \Leftrightarrow [\![\varphi(\bar{a}, A_1, \ldots, A_k)]\!] \in \mathbb{G}.$$

The proof is identical to Theorem 4.1.

Theorem 5.2. *Let* $\mathfrak{M} \models \mathbf{VL}$. *If* $\mathbb{G} \subseteq \mathbb{B}_{BP}$ *is TY-generic then* $\mathfrak{M}[\mathbb{G}] \models \mathbf{VL}$.

First we prove two technical lemmas.

Lemma 5.2. *Let* $\varphi(\bar{x}, \bar{X}, Y) \in \Sigma_0^B$ *and* $t(\bar{x}, \bar{n})$ *be a* $\mathcal{L}_A^2$ *term and suppose that*

$$\mathbf{VL} \vdash \forall \bar{x}, \bar{X} \exists! Y \leq t(\bar{x}, |\bar{X}|) \; \varphi(\bar{x}, \bar{X}, Y).$$

Then $\mathbf{VL}$ *proves that the unique* Y *such that* $\varphi(\bar{x}, \bar{X}, Y)$ *is bitwise computable by branching programs:*

$\forall \bar{x} \forall \bar{n} \exists Z \; [Seq(Z) \wedge Len(Z) \leq t(\bar{x}, \bar{n}) \wedge \forall i < Len(Z) BP(n_i, (Z)_i) \wedge$
$\forall \bar{X} : |\bar{X}| = \bar{n} \forall Y < t(\bar{x}, \bar{n})$
$(|Y| = Len(Z) \wedge \forall i < t(\bar{x}, \bar{n})(Y(i) \leftrightarrow \bar{X} \models_{BP} (Z)_i) \to \varphi(\bar{x}, \bar{X}, Y))].$

Proof. Let $\overline{\mathbf{VL}}$ be the universal conservative extension of $\mathbf{VL}$ as defined in Cook and Nguyen.[6] Then for any $\varphi(\bar{x}, \bar{X}, Y) \in \Sigma_0^B$ and $t(\bar{x}, \bar{n})$ such that

$$\mathbf{VL} \vdash \forall \bar{x} \bar{X} \exists! Y \leq t(\bar{x}, |\bar{X}|) \varphi(\bar{x}, \bar{X}, Y)$$

there exists a function symbol $F_{\varphi, t}(\bar{x}, \bar{X})$ of $\overline{\mathbf{VL}}$ such that

$$\overline{\mathbf{VL}} \vdash \forall \bar{x} \bar{X} \varphi(\bar{x}, \bar{X}, F_{\varphi, t}(\bar{x}, \bar{X})) \wedge |F_{\varphi, t}(\bar{x}, \bar{X}))| \leq t(\bar{x}, |\bar{X}|).$$

For each function symbol $F(\bar{x}, \bar{X}))$ of $\overline{\mathbf{VL}}$, $\overline{\mathbf{VL}}$ proves that

$\forall \bar{x} \forall \bar{n} \exists Z \; [Seq(Z) \wedge Len(Z) \leq t(\bar{x}, \bar{n}) \wedge \forall i < Len(Z) BP(n_i, (Z)_i) \wedge$
$\forall \bar{X} : |\bar{X}| = \bar{n} \forall Y < t(\bar{x}, \bar{n})$
$(|Y| = Len(Z) \wedge \forall i < t(\bar{x}, \bar{n})(Y(i) \leftrightarrow \bar{X} \models_{BP} (Z)_i) \to F(\bar{x}, \bar{X}) = Y)].$

This is proved by induction on the complexity of F and the claim immediately follows from these two facts. $\square$

Lemma 5.3. *Let* $\varphi(\bar{x}, X_1, \ldots, X_k) \in \Sigma_0^B$ *and* $t_1 \ldots, t_k$ *be* $\mathcal{L}_A^2$ *terms.* $\mathbf{VL}$ *proves that*

$\forall \bar{a} \forall C_1, \ldots, C_k \forall V_1, \ldots, V_k \forall A \in 2^n$
$\bigwedge_{1 \leq i \leq k} \left\{ \begin{array}{l} Seq(C_i) \wedge Len(C_i) = |V_i| = t_i \wedge \forall j < t_i BP(n, (C_i)_j) \wedge \\ \forall j < t_i \; (V_i(j) \leftrightarrow A \models_{BP} (C_i)_j) \end{array} \right\}$
$\to ((\langle V_1, \ldots, V_k \rangle \models_A \|\varphi(\bar{x}, \bar{X})\| \leftrightarrow A \models_{BP} [\![\varphi(\bar{a}, C_1, \ldots, C_k)]\!]).$

The proof is straightforward and we omit the details.

Proof of Theorem 5.2. By Theorem 5.1, it suffices to show that $\mathfrak{M}[\mathbb{G}] \models PATH$. We actually show that for any $a \in M_0$ and $\hat{E} : \langle a, a \rangle \to \mathbb{B}_{BP}$ there exists $\hat{P} : \langle a, a \rangle \to \mathbb{B}_{BP}$ such that

$$[\![Unique(a, \hat{E}) \to \delta_{PATH}(a, \hat{E}, \hat{P})]\!] =_{BP} 1.$$

By Lemma 5.2 we have a sequence of branching programs

$$Z = \langle Z_0, \ldots, Z_{\langle a,a \rangle - 1} \rangle \in M$$

which witnesses P such that

$$Unique(a, \hat{E}) \to \delta_{PATH}(a, \hat{E}, P).$$

For arbitrary $\hat{E} : \langle a, a \rangle \to \mathbb{B}_{BP}$ define

$$\hat{P} : \langle a, a \rangle \to \mathbb{B}_{BP}, \hat{P}(i) = Z_i(\hat{E}).$$

Let $A \in 2^n$ be arbitrary. By Lemma 5.3 there exist $\tilde{E}, \tilde{P} \in M$ such that

$$|\tilde{E}| \leq \langle a, a \rangle \wedge \forall i < \langle a, a \rangle (\tilde{E} \leftrightarrow A \models_{BP} \hat{E}) \wedge$$
$$|\tilde{P}| \leq \langle a, a \rangle \wedge \forall i < \langle a, a \rangle (\tilde{P} \leftrightarrow A \models_{BP} \hat{P}) \wedge$$
$$\langle \tilde{E}, \tilde{P} \rangle \models_A \|Unique(x, E) \to \delta_{PATH}(x, E, P)\|_{a,\langle a,a \rangle, \langle a,a \rangle}$$
$$\leftrightarrow A \models_{BP} [\![Unique(x, E) \to \delta_{PATH}(x, E, P)]\!]$$

is true in $\mathfrak{M}$.

On the other hand, we have

$$\mathfrak{M} \models \langle \tilde{E}, \tilde{P} \rangle \models_A \|Unique(a, E) \to \delta_{PATH}(a, E, P)\|_{a,\langle a,a \rangle, \langle a,a \rangle}.$$

Hence we conclude that

$$\mathfrak{M} \models [\![Unique(a, \hat{E}) \to \delta_{PATH}(a, \hat{E}, \hat{P})]\!] =_{BP} 1$$

which proves the claim of the theorem. $\square$

Remark 5.1. We do not define Boolean algebra for L based on propositional proof system. In Buss, Das and Knop,[4] a propositional proof system based on branching programs is defined which will be a candidate for the right system to define such a Boolean algebra. However, it is still open whether the system corresponds to **VL**.

6. Boolean valued models for NL

In order to construct Boolean algebras for NL, we define a circuit model which allows to use connectives for the transitive closure. Transitive closure operators are used to define various complexity classes on finite structures and we borrow ideas from such characterizations. For details about transitive closure and complexity classes, see Immermann.[7]

Definition 6.1. TC formulas are defined inductively as follows:

- constants 0, 1 and propositional variables are TC formulas,
- if ϕ_0 and ϕ_1 are TC formulas then so are $\phi_0 \wedge \phi_1$, $\phi_0 \vee \phi_1$ and $\neg \phi_0$,
- if $\phi_{i,j}$ are TC formulas for $i, j < n$ then so is

$$TC^k_{n,a,b}(\bar{\phi}_0, \ldots, \bar{\phi}_{n-1})$$

for $a, b, k < n$ where $\bar{\phi}_i = \phi_{i,0}, \ldots, \phi_{i,i-1}, \phi_{i,i+1}, \ldots, \phi_{i,n-1}$ is a list of TC formulas for $0 \leq i < n$.

Let $\bar{p} = p_0, \ldots, p_{n-1}$ be a list of propositional formulas. Then $TC(\bar{p})$ denotes the set of TC formulas with variables among $\bar{p}$.

Intended meaning of the connective $TC^k_{n,a,b}$ is that it holds if there exists a path form a to b of length less than or equal to k in the graph with n vertices in which there is an edge from i to j if and only if $\phi_{i,j}$ is true.

Definition 6.2. For a TC formula $\phi(\bar{p})$ with variables among $\bar{p}$ and assignments $A = a_0 \cdots a_{n-1} \in 2^n$, we define the satisfaction relation $A \models_{TC} \phi(\bar{p})$ by extending the satisfaction relation for propositional formulas by the following rule:

- $A \models_{TC} TC^0_{n,a,a}(\bar{\phi}_0, \ldots, \bar{\phi}_{n-1})$.
- $A \models_{TC} TC^0_{n,a,b}(\bar{\phi}_0, \ldots, \bar{\phi}_{n-1})$ if $a \neq b$.
- For $k < n - 1$, $A \models_{TC} TC^{k+1}_{n,a,b}(\bar{\phi}_0, \ldots, \bar{\phi}_{n-1})$ if there exists $c < n$ such that $A \models_{TC} TC^k_{n,a,c}(\bar{\phi}_0, \ldots, \bar{\phi}_{n-1})$ and $A \models_{TC} \phi_{c,b}$.

For $\phi, \psi \in TC(\bar{p})$, $\phi \leq_{TC} \psi$ if

$$A \models_{TC} \phi \to A \models_{TC} \psi$$

for all $A \in 2^n$. $TC(\bar{p})$ is partially ordered with respect to $\leq_{TC}$. We will also give a partial order based on the propositional provability in $TC(\bar{p})$.

Definition 6.3. $TCPK$ is the system PK which operates on TC formulas and extended by the following extra axioms.

Axioms for TC connectives :

$$\to TC^0_{n,a,a}(\bar{\phi}_0, \ldots, \bar{\phi}_{n-1}).$$
$$TC^0_{n,a,b}(\bar{\phi}_0, \ldots, \bar{\phi}_{n-1}) \to 0 \text{ if } a \neq b.$$
$$TC^{k+1}_{n,a,b}(\bar{\phi}_0, \ldots, \bar{\phi}_{n-1}) \leftrightarrow \bigvee_{c<n} (TC^k_{n,a,c}(\bar{\phi}_0, \ldots, \bar{\phi}_{n-1}) \wedge \phi_{c,b}).$$

Using this proof system, we can define another partial order on $TC(\bar{p})$

$$\phi \leq_{TCPK} \psi \Leftrightarrow \text{ there exists an TCPK-proof of } \psi \text{ from } \phi.$$

It is easy to see that $TCPK$ is a sound and complete propositional proof system. We will show that $TCPK$ is a right system for NL reasonings.

More precisely, there exists a bounded arithmetic theory for NL such that

- all Σ^B_0 theorems of the theory are translated into TC tautologies which have polynomial size $TCPK$ proofs and
- the theory proves the reflection principle for $TCPK$.

Definition 6.4. Let $\mathcal{L}_{TC}$ be the language $\mathcal{L}^2_A$ extended by introducing the following predicate symbols: for $\varphi(x, y, \bar{z}, \bar{X}) \in \Sigma^B_0(\mathcal{L}_{TC})$, we introduce a new predicate symbol

$$TC_{x,y}[\varphi(x, y, \bar{z}, \bar{X})](a, b, k, n, \bar{z}, \bar{X})$$

with variables as displayed.

Definition 6.5. $\mathbf{V}^0(TC)$ is the $\mathcal{L}_{TC}$ theory whose axioms are BASIC axioms, $\Sigma^B_0(\mathcal{L}_{TC})$-COMP and the following defining axioms for

TC predicates:

$$a < n \to TC_{x,y}[\varphi(x, y, \bar{z}, \bar{X})](a, a, 0, n, \bar{z}, \bar{X}),$$
$$a, b < n \land a \neq b \to \neg TC_{x,y}[\varphi(x, y, \bar{z}, \bar{X})](a, b, 0, n, \bar{z}, \bar{X}),$$
$$a, b < n \land k < n - 1 \to$$
$$TC_{x,y}[\varphi(x, y, \bar{z}, \bar{X})](a, b, k + 1, n, \bar{z}, \bar{X})$$
$$\leftrightarrow \exists c < n(TC_{x,y}[\varphi(x, y, \bar{z}, \bar{X})](a, c, k, n, \bar{z}, \bar{X}) \land \varphi(c, b, \bar{z}, \bar{X})),$$
$$n \leq a \lor n \leq b \lor n \leq k \to \neg TC_{x,y}[\varphi(x, y, \bar{z}, \bar{X})](a, b, k, n, \bar{z}, \bar{X}).$$

Theorem 6.1. $\mathbf{V}^0(TC)$ *and* $\mathbf{VNL}$ *proves the same* $\mathcal{L}_A^2$ *theorems.*

Proof. First we show that $\mathbf{V}^0(TC)$ proves

$$\forall a \forall E < \langle a, a \rangle \exists Y < \langle a, a \rangle\, \delta_{CONN}(a, E, Y).$$

Define

$$\varphi(a, b, c, E) \equiv b < a \land c < a \land E(b, c)$$

and

$$\phi(a, z, x, E) \equiv TC_{b,c}[\varphi(a, b, c, E)](a, z, 0, x, E).$$

By Σ_0^B-COMP in $\mathbf{V}^0(TC)$ we have Y such that

$$\forall z, x < a\ (Y(z, x) \leftrightarrow \psi(a, z, x, E))$$

and Using TC axioms we can show that $\delta_{CONN}(a, E, Y)$ provably in $\mathbf{V}^0(TC)$.

For the converse inclusion, we will show that TC predicates are Δ_1^B definable in $\mathbf{VNL}$. Since $\mathbf{VNL}$ Δ_1^B-defines all NL predicates and TC predicates represents the reachability problem which is in NL, this is straightforward.

As we can conservatively add Δ_1^B definable predicates to $\mathbf{VNL}$ we have the claim. $\square$

We will construct generic extensions based on two Boolean algebras with $TC(\bar{p})$ as their underlying set.

We can naturally extend the propositional translation to $\Sigma_0^B(\mathcal{L}_{TC})$ formulas.

Definition 6.6. For $\varphi(\bar{x}, \bar{X}) \in \Sigma_0^B(\mathcal{L}_{TC})$, we define its propositional translation by adding the following rule:

$$\|TC_{x,y}[\varphi(x, y, \bar{z}, \bar{X})](a, b, k, n, \bar{z}, \bar{X})\|_{a,b,k,n,\bar{l},\bar{m}}\bar{x}_0, \ldots, \bar{x}_{m-1}$$
$$:= TC_{n,a,b}^k(\|\bar{\varphi}\|)$$

where

$$\|\bar{\varphi}\| := \psi_0, \ldots, \psi_{n-1},$$
$$\psi_i := \phi_{i,0}, \ldots, \phi_{i,i-1}, \phi_{i,i+1}, \ldots, \phi_{i,n-1} \text{ for } 0 \le i < n,$$
$$\phi_{i,i} := 1,$$
$$\phi_{i,j} := \|\varphi(x, y, \bar{z}, \bar{X})\|_{i,j,\bar{l},\bar{m}}(\bar{x}_0, \ldots, \bar{x}_{m-1}) \text{ for } i \ne j.$$

Theorem 6.2. Let $\varphi(\bar{x}, \bar{X}) \in \Sigma_0^B(\mathcal{L}_{TC})$. If $\mathbf{V}^0(TC) \vdash \forall \bar{x} \forall \bar{X} \varphi(\bar{x}, \bar{X})$ then $\|\varphi(\bar{x}, \bar{X})\|_{\bar{l},\bar{m}}$ have $TCPK$ proofs whose size are polynomial in $\bar{l}$ and $\bar{m}$.

Proof. It suffices to show that the propositional translation of the axioms for TC predicates have polynomial size $TCPK$ proof which is obvious. $\square$

Theorem 6.3. $\mathbf{V}^0(TC)$ proves that $TCPK$ is sound.

Proof. It suffices to show that the predicate

$$A \models X \Leftrightarrow TCF(X) \to A \text{ is a satisfying assignnment for } X$$

is Δ_1^B-definable in $\mathbf{V}^0(TC)$. This follows from the fact that the reachability problem is in NL and any predicate in NL is Δ_1^B-definable in $\mathbf{V}^0(TC)$. $\square$

Remark 6.1. We conjecture that $TCPK$ and G_{NL}^* are p-equivalent. But we do not discuss about this problem here.

We consider two Boolean algebras with $TC(\bar{p})$ as their underlying set.

Definition 6.7. Define

$$\mathbb{B}_{TC} = TC(\bar{p})/ =_{TC}, \quad \mathbb{B}_{TCPK} = TC(\bar{p})/ =_{TCPK}.$$

The following theorem is proved in the same way as for NC^1.

Theorem 6.4 (Forcing Theorem for NL). *Let* $\mathfrak{M} \models \mathbf{VNL}$, $\mathbb{B}$ *be either* $\mathbb{B}_{TC}$ *or* $\mathbb{B}_{TCPK}$, $\varphi(\bar{x}, X_1, \ldots, X_k) \in \Sigma_0^B$, $\bar{a} \in M_0$ *and* $A_1, \ldots, A_k \in M^{\mathbb{B}}$. *Then for any TY-generic* $\mathbb{G} \subseteq \mathbb{B}$

$$\mathfrak{M}[\mathbb{G}] \models \varphi(\bar{a}, i_{\mathbb{G}}(A_1), \ldots, i_{\mathbb{G}}(A_k)) \Leftrightarrow [\![\varphi(\bar{a}, A_1, \ldots, A_k)]\!] \in \mathbb{G}.$$

Theorem 6.5. *Let* $\mathfrak{M} \models \mathbf{VNL}$ *and* $\mathbb{B}$ *be either* $\mathbb{B}_{TC}$ *or* $\mathbb{B}_{TCPK}$. *If* $\mathbb{G} \subseteq \mathbb{B}$ *is a TY-generic then* $\mathfrak{M}[\mathbb{G}] \models \mathbf{VNL}$.

Proof. Let $\mathfrak{M} \models \mathbf{VNL}$ and $\mathbb{B}$ be either $\mathbb{B}_{TC}$ or $\mathbb{B}_{TCPK}$. By Theorem 6.4 any generic extension satisfies $\mathbf{V}^0$. Hence it suffices to show that $\mathfrak{M}[\mathbb{G}] \models CONN$ whenever $\mathbb{G} \subseteq \mathbb{B}$ is a TY-generic. Remark that

$$[\![\delta_{CONN}(a, E, Y)]\!] =$$
$$Y(0,0) \wedge \bigwedge_{x<a, x\neq 0} \neg Y(0,x) \wedge$$
$$\bigwedge_{z<a-1} \bigwedge_{x<a} \left\{ Y(z+1, x) \leftrightarrow \left(Y(z,x) \vee \bigvee_{y<a} (Y(z,y) \wedge E(y,x)) \right) \right\}.$$

Using TC connectives we define $Y : \langle a, a \rangle \to \mathbb{B}$ as

$$Y(0,0) = 1,$$
$$Y(0,x) = 0 \text{ for } x \neq 0, x \leq a,$$
$$Y(z,x) = TC_{a,0,x}^k(\bar{\phi}_0, \ldots, \bar{\phi}_{a-1})$$

where $\phi_i = \phi_{i,0}, \ldots, \phi_{i,i-1}, \phi_{i,i+1}, \ldots, \phi_{i,a-1}$ with $\phi_{i,j} = E(i,j)$. Then it is easily seen that

$$TCPK \vdash [\![\delta_{CONN}(a, E, Y)]\!].$$

So $[\![\delta_{CONN}(a, E, Y)]\!]. =_{TCPK} 1$. By Theorem 6.3, we have

$$[\![\delta_{CONN}(a, E, Y)]\!]. =_{TC} 1$$

Thus $[\![\delta_{CONN}(a, E, Y)]\!]. =_{TC} 1 \in \mathbb{G}$ where $\mathbb{G} \subseteq \mathbb{B}$ is a TY-generic. $\square$

7. Separation problems and generic extensions

We will establish some connections between separation problems of computational and proof complexity in the ground model and generic extensions.

First we show that if $P = NP$ in the ground model then the generic extension satisfies Σ_1^B induction which was originally proved in Takeuti and Yasumoto.[13]

Theorem 7.1. *Let* $\mathfrak{M} \models \mathbf{VP}$. *If* $\mathfrak{M} \models (P = NP)$ *then* $\mathfrak{M}[\mathbb{G}] \models \mathbf{V}^1$ *for any TY-generic* $\mathbb{G} \subseteq \mathbb{B}_C$.

Proof. Let $\mathfrak{M} \models \mathbf{VP}$ and suppose that $\mathfrak{M} \models (P = NP)$. Let $\varphi(x, \bar{y}, X_1, \ldots, X_k, Z) \in \Sigma_0^B$ and $t(m_1, \ldots, m_k)$ be a term, $a, \bar{b} \in M_0$ and $A_1, \ldots, A_k \in M_C^{\mathbb{B}}$ with $A_i : m_i \to \mathbb{B}_C$.

We will show that

$$\mathfrak{M}[\mathbb{G}] \models \exists Y < a \forall x < a \ (Y(x) \leftrightarrow \exists Z \leq t(\bar{m}) \varphi(x, \bar{y}, (A_1)_{\mathbb{G}}, \ldots, (A_k)_{\mathbb{G}}, Z)).$$

Since $\mathfrak{M} \models (P = NP)$, SAT is computed by circuits in $\mathfrak{M}$. Therefore, for each $i < a$ there is a circuit $C_i(\bar{q}_1, \ldots, \bar{q}_k)$ such that

$\forall \alpha_1 \in 2^{m_1} \cdots \forall \alpha_k \in 2^{m_k}$
$$\left(\begin{array}{l} \langle \alpha_1, \ldots, \alpha_k \rangle \models_A C_i(\bar{q}_1, \ldots, \bar{q}_k) \leftrightarrow \\ \exists \beta \in 2^t \langle \alpha_1, \ldots, \alpha_k \rangle \models_A \|\varphi(x, \bar{y}, X_1, \ldots, X_k, Z)\|_{i, \bar{b}, \bar{m}, t}(\bar{q}_1, \ldots, \bar{q}_k, \bar{z}) \end{array} \right).$$

holds in $\mathfrak{M}$.

Using binary search, we can construct circuits

$$D_{i,0}(\bar{q}_1, \ldots, \bar{q}_k), \ldots, D_{i,t-1}(\bar{q}_1, \ldots, \bar{q}_k)$$

such that

$\forall \alpha_1 \in 2^{m_1} \cdots \forall \alpha_k \in 2^{m_k}$
$$\left(\begin{array}{l} \langle \alpha_1, \ldots, \alpha_k \rangle \models_A \|\varphi(x, \bar{y}, X_1, \ldots, X_k, Z)\|_{i, \bar{b}, \bar{m}, t}(\bar{q}_1, \ldots, \bar{q}_k, \bar{D}_i(\bar{q}_1, \ldots, \bar{q}_k)) \\ \leftrightarrow \exists \beta \in 2^t \langle \alpha_1, \ldots, \alpha_k \rangle \models_A \|\varphi(x, \bar{y}, X_1, \ldots, X_k, Z)\|_{i, \bar{b}, \bar{m}, t}(\bar{q}_1, \ldots, \bar{q}_k, \bar{z}) \end{array} \right)$$

holds in $\mathfrak{M}$.

Define $Y : a \to \mathbb{B}_C$ by $Y(i) = C_i(A_1, \ldots, A_k)$ for $i < a$ and for each $i < a$ define $B_i : t \to \mathbb{B}_C$ by

$$B_i(j) = D_{i,j}(A_1, \ldots, A_k)$$

for $k < t$.

Then it is easy to see that

$$Y(i) \leq_A [\![\varphi(i, \bar{b}, A_1, \ldots, A_k, B_i)]\!]$$

and

$$[\![\varphi(i, \bar{b}, A_1, \ldots, A_k, Z)]\!] \leq_A Y(i)$$

for any $Z : t \to \mathbb{B}_C$. $\square$

Theorem 7.2. *Let* $\langle \mathcal{C}, \mathbb{B} \rangle$ *be either* $\langle NC^1, \mathbb{B}_F \rangle$, $\langle L, \mathbb{B}_{BP} \rangle$ *or* $\langle NL, \mathbb{B}_{TC} \rangle$. *Let* $\mathfrak{M} \models \mathbf{VP}$ *and suppose that* $\mathfrak{M} \models (P \subseteq \mathcal{C})$. *Then* $\mathfrak{M}[\mathbb{G}] \models \mathbf{VP}$ *for any TY-generic* $\mathbb{G} \subseteq \mathbb{B}$.

Proof. If $\mathfrak{M} \models (P = NC^1)$ then $\mathbb{B}_F = \mathbb{B}_C$. So the claim follows form Theorem 3.6. $\square$

The converse of Theorem 7.2 also holds with the change of Boolean algebra.

Theorem 7.3. *Let* $\langle \mathcal{C}, \mathbb{B} \rangle$ *be either* $\langle NC^1, \mathbb{B}_{G_0} \rangle$, *or* $\langle NL, \mathbb{B}_{TCPK} \rangle$. *Let* $\mathfrak{M} \models \mathbf{V}^1$ *and suppose that* $\mathfrak{M} \models (P \not\subseteq \mathcal{C})$. *Then there exists a TY-generic* $\mathbb{G} \subseteq \mathbb{B}$ *such that*

$$\mathfrak{M}[\mathbb{G}] \not\models \mathbf{VP}.$$

To prove Theorem 7.3, we use the idea similar to the one used in Krajíček.[10]

Proof. We prove the case for the pair $\langle NC^1, \mathbb{B}_{G_0} \rangle$. Suppose that $\mathfrak{M} \models P \not\subseteq NC^1$. Let $(\bar{p})_0$ and $(\bar{p})_1$ be the lists of variables in $\bar{p}$ such that

$$(\bar{p})_0 = p_0, \ldots, p_{a-1}, (\bar{p})_1 = p_a, \ldots, p_{\langle a, a \rangle - 1}$$

where a is the maximal number such that $a + \langle a, a \rangle \le n$. Consider the set

$$\mathbb{S} = \{\neg[\![\delta_{MCV}(a, (\bar{p})_0, (\bar{p})_1, Y)]\!] : Y : a + 2 \to \mathbb{B}_{G_0}\}.$$

We claim that there is no G_0-proof of $\bot$ from $\mathbb{S}$. If otherwise, there exists $Y_0, \ldots, Y_k : a + 1 \to \mathbb{B}_{G_0}$ such that

$$\bigwedge_{i \le k} (\neg[\![\delta_{MCV}(a, (\bar{p})_0, (\bar{p})_1, Y_i)]\!] \to \bot)$$

has a G_0-proof. Since $\mathbf{V}^1$ proves the soundness of G_0,

$$\bigvee_{i \le k} [\![\delta_{MCV}(a, (\bar{p})_0, (\bar{p})_1, Y_i)]\!]$$

is a tautology in $\mathfrak{M}$ and using this formula, we can construct an NC^1 algorithm deciding MCV which contradicts to the assumption.

So $\mathbb{S}$ is G_0-consistent in $\mathfrak{M}$. Since Theorem 3.8 holds for G_0 instead of CF, there exists a TY-generic $\mathbb{G} \subseteq \mathbb{B}_{G_0}$ such that $\mathbb{S} \subseteq \mathbb{G}$ and so

$$\mathfrak{M}[\mathbb{G}] \models \neg\varphi(a, i_{\mathbb{G}}((\bar{p})_0), i_{\mathbb{G}}((\bar{p})_1), Y)$$

for any $Y : a + 2 \to \mathbb{B}_{G_0}$ which proves the theorem. $\qquad\square$

We can also establish connections with properties of propositional proof systems.

Theorem 7.4. *Let $\mathfrak{M} \models \mathbf{V}^1$ and suppose that the following conditions hold in $\mathfrak{M}$.*

- $G_1^* \leq_p G_0$.
- *any Σ_1^B-theorem of $\mathbf{V}^1$ has polynomial size G_1^*-proofs.*

Then $\mathfrak{M}[\mathbb{G}] \models \mathbf{VP}$ for any TY-generic $\mathbb{G} \subseteq \mathbb{B}_{G_0}$. This also holds for G_{NL}^* and $\mathbb{B}_{TCPK}$ instead of G_0 and $\mathbb{B}_{G_0}$ respectively.

Proof. It suffices to show that $\mathfrak{M}[\mathbb{G}] \models MCV$. By translation theorem, there exist G_1^*-proofs of the following formula for all a provably in $\mathbf{VP}$:

$$\exists y \| \delta_{MCV}(a, G, E, Y) \| (\bar{g}, \bar{e}, \bar{y}) \quad (\dagger)$$

where $\bar{g}, \bar{e}, \bar{y}$ are propositional variables representing G, E and Y respectively such that $|G| < a$, $|E| < \langle a, a \rangle$ and $|Y| < a + 2$. Since $\mathfrak{M} \models G_1^* \leq_p G_0$, there exists a G_0 proof of $(\dagger)$ and by the second assumption, there exists an NC^1 function $F(P, \bar{g}, \bar{e})$ such that if P is a G_0-proof of $(\dagger)$ then $F(P, \bar{g}, \bar{e})$ witnesses $\bar{y}$ for inputs $\bar{g}$ and $\bar{e}$. Moreover, F can be represented by a sequence of propositional formulas

$$F_0(P, \bar{g}, \bar{e}), \ldots, F_{a+1}(P, \bar{g}, \bar{e}).$$

Fix a proof $P \in M$ of $(\dagger)$ and define

$$Y : a + 2 \to \mathbb{B}_{G_0}, Y(i) = F_i(P, G, E).$$

Then it is readily seen that

$$[\![\delta_{MCV}(a, G, E, Y)]\!] = \exists \bar{y} \, \| \delta_{MCV}(a, G, E, Y) \| (G, E, Y)$$

and

$$\mathfrak{M} \models [\![\delta_{MCV}(a, G, E, Y)]\!] =_F 1.$$

The claim of the theorem follows from this and the forcing theorem.

□

It is unknown whether the converse holds. Nevertheless, by assuming the unprovability of some principle in G_0, we can construct a generic extension in which its nagation is true.

Theorem 7.5. Let $\mathfrak{M} \models \mathbf{V}^1$ and $\tau(\bar{p}, y_0, \ldots, y_{m-1}) \in M$ be a propositional formula with variables among $\bar{p}, y_0, \ldots, y_{m-1}$. If $\mathfrak{M} \models (G_0 \nvdash \exists \bar{y} \tau(\bar{p}, \bar{y}))$ then there exists a TY-generic $\mathbb{G} \subseteq \mathbb{B}_{G_0}$ such that

$$\{\neg \tau(\bar{p}, Y(0), \ldots, Y(m-1)) : Y : m \to \mathbb{B}_{G_0}\} \subseteq \mathbb{G}.$$

Proof. Let $\mathfrak{M}$ and τ be as above. We claim that

$$\{\neg \tau(\bar{p}, Y(0), \ldots, Y(m-1)) : Y : m \to \mathbb{B}_{G_0}\}$$

is G_0-consistent. To show this, suppose otherwise. Then there exist $l \in M_0$ and $Y_0, \ldots, Y_{l-1} : m \to \mathbb{B}_{G_0}$ from which 0 is G_0-provable in $\mathfrak{M}$. Therefore we have

$$G_0 \vdash \bigvee_{i < l} \tau(\bar{p}, Y_i(0), \ldots, Y_i(m-1))$$

and from such a proof we construct a G_0-proof of $\exists \bar{y} \tau(\bar{p}, \bar{y})$ which is a contradiction.

Hence we have a TY-generic $\mathbb{G} \subseteq \mathbb{B}_{G_0}$ such that

$$\{\neg \tau(\bar{p}, Y(0), \ldots, Y(m-1)) : Y : m \to \mathbb{B}_{G_0}\} \subseteq \mathbb{G}. \qquad \square$$

In particular we have

Corollary 7.1. Let $\mathfrak{M} \models \mathbf{V}^1$ and suppose that in $\mathfrak{M}$ there is no G_0-proof of the propositional statement that any circuit can be evaluated. Then there exists a TY-generic $\mathbb{G} \subseteq \mathbb{B}_{G_0}$ such that

$$\mathfrak{M}[\mathbb{G}] \not\models \mathbf{VP}.$$

Proof. Let $\tau \equiv [\![\delta_{MCV}(a, G, E, Y)]\!]$ and apply Theorem 7.5. □

8. Σ_2^B and Π_2^B formulas in generic extensions

Finally, we consider the problem of how formulas of higher logical complexity perform in the generic extension. In particular we state and prove three results concerning Σ_2^B and Π_2^B formulas. These results are presented for the Boolean algebra for PTIME and the propositional proof system G_1^*. We remark that our results in this section hold for other complexity classes and propositional proof systems with slight modifications.

Theorem 8.1. $\mathfrak{M} \models \mathbf{V}^1$, $\varphi(X, Y) \in \Sigma_0^B$, $t(x)$ be a term and $a \in M_0$. If $\mathfrak{M} \models (G_1^* \vdash \exists \bar{y} \, \|\varphi(X, Y)\|_{a,t(a)}(\bar{x}, \bar{y}))$ then

$$\mathfrak{M}[\mathbb{G}] \models \forall X < a \, \exists Y < t(a) \, \varphi(X, Y)$$

for any TY-generic $\mathbb{G} \subseteq \mathbb{B}_C$.

Proof. Let $\varphi(X, Y) \in \Sigma_0^B$ and suppose that

$$\mathfrak{M} \models (G_1^* \vdash \exists \bar{y} \| \varphi(X, Y) \|_{a,t(a)}(\bar{x}, \bar{y})).$$

Since $\Sigma_1^q\text{-}Wit_{G_1^*}$ is in PTIME (Cook-Nguyen[6]), there exists a polynomial time function $F(X)$ such that

$$\mathfrak{M} \models \forall X \in 2^n(\langle X, F(X) \rangle \models \|\varphi(X, Y)\|_{a,t(a)}(\bar{x}, \bar{y})).$$

Let $C_0, \ldots, C_{t(a)-1}$ be a list of Boolean circuits representing $F(X)$ for $|X| = a$ and for $X : a \to \mathbb{B}_C$, define $Y : t(a) \to \mathbb{B}_C$ by $Y[i] = C_i(X)$ for $i < t(a)$. Then it is easy to see that $[\![\varphi(X, Y)]\!] =_C 1$ which proves the theorem. $\qquad \square$

Theorem 8.2. Let $\mathfrak{M} \models \mathbf{V}^2$, $\varphi(X, Y, Z) \in \Sigma_0^B(PV)$, $t(a)$ be a term and $a, b \in M_0$. Suppose that $\mathfrak{M} \models (P = NP)$. If

$$\mathfrak{M} \models (G_1^* \vdash \exists \bar{x} \, \forall \bar{y} \, \|\varphi(X, Y, Z)\|_{a,t(a),b}(\bar{x}, \bar{y}, \bar{z}))$$

then

$$\mathfrak{M}[\mathbb{G}] \models \forall Z < b \, \exists X < a \, \forall Y < t(a) \, \varphi(X, Y, Z)$$

for any TY-generic $\mathbb{G} \subseteq \mathbb{B}_C$.

To prove the theorem, we need the following technical fact:

Lemma 8.1. *There exists a FP^{NP} function $F(P, Z)$ which is Σ_2^B definable in $\mathbf{V}^2$ such that*

$$\forall \phi \, \forall P \, (Prf_{G_1^*}(P, \exists \bar{x} \forall \bar{y} \phi(\bar{x}, \bar{y}, \bar{z})) \to \forall Z \forall Y \langle F(P, Z), Y, Z \rangle \models_F \phi(\bar{x}, \bar{y}, \bar{z}))$$

is provable in $\mathbf{V}^2$.

Proof. This is implied by the fact that the Σ_2^q witnessing problem for G_1^* is in FP^{NP} and that $\mathbf{V}^2$ Σ_2^B-defines all FP^{NP} functions. $\quad\square$

Proof of Theorem 8.2. Let $\varphi(X, Y, Z) \in \Sigma_0^B(PV)$, t be a term, $a, b \in M_0$ and $P \in M$ be such that

$$\mathfrak{M} \models Prf_{G_1^*}(P, \exists \bar{x} \forall \bar{y} \| \varphi(X, Y, Z) \|_{a, t(a), b}(\bar{x}, \bar{y}, \bar{z})).$$

Then by Lemma 8.1, there is a FP^{NP} function $F(P, Z)$ witnessing the existential quantifiers $\exists \bar{x}$ in the propositional translation of

$$\forall Z < b \, \exists X < a \, \forall Y < t(a) \, \varphi(X, Y, Z).$$

Assume that $\mathfrak{M} \models (P = NP)$. Then $F(P, Z) \in FP$. Let $C_0, \ldots, C_{a-1}$ be a list of Boolean circuits representing $F(P, X)$ for $\| \varphi(X, Y, Z) \|_{a, t(a), b}(\bar{x}, \bar{y}, \bar{z})$. Let $Z : b \to \mathbb{B}_C$ and define $X : a \to \mathbb{B}_C$ as $X[i] = C_i(Z)$. Then we have $[\![\varphi(X, Y, Z)]\!] =_C 1$ for any $Y : t(a) \to \mathbb{B}_C$ which proves the theorem. $\quad\square$

As an application of Theorem 8.2, consider the dual weak pigeonhole principle

$$dWPHP_a^{a+1}(F) \equiv \exists Y < a + 1 \, \forall X < a \, (F(X) \neq Y)$$

where F is a PV function.

Let $\mathfrak{M} \models \mathbf{V}^2$ and suppose that there exists a TY-generic $\mathbb{G} \subseteq \mathbb{B}_C$ such that

$$\mathfrak{M}[\mathbb{G}] \models \neg dWPHP(F)_a^{a+1}.$$

Then by Theorem 8.2, either

(1) $\mathfrak{M} \models (P \neq NP)$ or
(2) $\mathfrak{M} \models (G_1^* \not\vdash \exists \bar{y} \, \forall \bar{x} \, \| F(X) \neq Y \|_{a, a+1}(\bar{x}, \bar{y}))$.

Jeřábek[8] defined a propositional proof system WF which extends CF by a special axiom representing the dual weak pigeonhole principle. Specifically, a WF-proof of a circuit A is a sequence of circuits $A_1, \ldots, A_k$ such that $A_k = A$ and each A_i with $1 \leq i \leq k$ is either an axiom of CF, obtained from $A_{j_1}, \ldots, A_{j_l}$ $j_1, \ldots, j_l < i$ by a Frege rule or a special axiom

$$\bigvee_{1 \leq l \leq m} (r_l \nleftrightarrow C_{i,l}(D_{i,1}, \ldots, D_{i,n}))$$

where $n < m$ and $r_1, \ldots, r_m$ are distinct variables which do not occur in $A, C_{i,1}, \ldots, C_{i,l}$ or $A_{j'}$ for $j' < i$ but may occur in $D_{i,1}, \ldots, D_{i,l}$.

It is easy to see that the condition (2) implies that G_1^* p-simulates WF. Hence we obtain

Corollary 8.1. *If there is a model* $\mathfrak{M} \models \mathbf{V}^2$ *and a TY-generic* $\mathbb{G} \subseteq \mathbb{B}_C$ *such that* $\mathfrak{M}[\mathbb{G}] \models \neg dWPHP(F)_a^{a+1}$ *then at least one of the following statements is consistent with* $\mathbf{V}^2$:

- $P \neq NP$,
- G_1^* *does not p-simulate WF.*

Theorem 8.3. *Let* $\mathfrak{M} \models \mathbf{V}^1$, $\varphi(X, Y) \in \Sigma_0^B$ *and* $t(x)$ *be a term. If*

$$\mathfrak{M} \models (G_1^* \nvdash \exists \bar{y} \, \|\varphi(X, Y)\|_{a, t(a)}(\bar{x}, \bar{y}))$$

then there exists a TY-generic $\mathbb{G} \subseteq \mathbb{B}_C$ *such that*

$$\mathfrak{M}[\mathbb{G}] \models \exists X < n \, \forall Y < t(n) \, \neg\varphi(X, Y).$$

Proof. Let $\varphi(X, Y) \in \Sigma_0^B$ and $t(x)$ be as above. Let $\alpha : n \to \mathbb{B}_C$ be defined by $\alpha[i] = p_i$ for $i < n$ and define

$$\mathbb{S} = \{\neg[\![\varphi(\alpha, Y)]\!] : Y : t(n) \to \mathbb{B}_C\}.$$

We claim that $\mathbb{S}$ is G_1^*-consistent in $\mathfrak{M}$.

If otherwise then there exist $k \in M_0$ and $Y_0, \ldots, Y_{k-1} : t(a) \to \mathbb{B}_C$ such that

$$\to \bigvee_{i < k} [\![\varphi(\alpha, Y_i)]\!]$$

is G_1^*-provable in $\mathfrak{M}$. It is also easy to see that

$$\bigvee_{i<k} [\![\varphi(\alpha, Y_i)]\!]$$

implies $\exists \bar{y} \ [\![\varphi(\alpha, Y_i)]\!]$ in G_1^* which is a contradiction. Hence by Theorem 3.8, there is a TY-generic $\mathbb{G} \subseteq \mathbb{B}_C$ such that $\mathbb{S} \subseteq \mathbb{G}$ which proves the theorem. $\square$

References

1. M. Ajtai, The complexity of the pigeonhole principle. Combinatorica, 14, pp.417-433, (1994).
2. A. Atserias and M. Müller, Partially definable forcing and bounded arithmetic. Archive for Mathematical Logic 54, pp.1-33, (2015).
3. S. R. Buss, Bounded Arithmetic. Ph.D Thesis, Princeton University. (1985).
4. S. Buss, A. Das, and A. Knop, Proof Complexity of (Non-deterministic) Decison Trees and Branching Programs. 28th EACSL Annual Conference on Computer Science Logic, 12:1–12:17, (2019).
5. S. A. Cook and T. Morioka, Quantified propositional calculus and a second-order theory for NC^1. Archive for Mathematical Logic, 44(6), pp.711-749, (2005).
6. S. A. Cook and P. Nguyen, Logical Foundations of Proof Complexity. Perspectives in Logic, Cambridge University Press. (2010).
7. N. Immerman, Descriptive Complexity. Springer. (1999).
8. E. Jeřábek, Dual weak pigeonhole principle, Boolean complexity, and derandomization. Annals of Pure and Applied Logic, 129, 1-3, pp.1-37, (2004).
9. J. Krajíček, On Frege and Extended Frege Proof Systems, in: Feasible Mathematics II, eds. P. Clote and J. Remmel, Birkhauser, pp.284-319, (1995).
10. J. Krajíček, Extensions of models of PV, in: Logic Colloquium'95, Eds. J.A. Makowsky and E.V. Ravve, ASL/Springer Series Lecture Notes in Logic, Vol. 11, pp.104-114 (1998),
11. J. Paris and A. J. Wilkie. Counting problems in bounded arithmetic. in: Methods in Mathematical Logic, 1130, pp.317-340, (1985).
12. S. Riis. Finitisation in bounded arithmetic. BRICS Report Series RS-94-23, (1994).
13. G. Takeuti and M. Yasumoto, Forcing on Bounded Arithmetic. in: Gödel '96, Lecture Notes in Logic, vol.6, pp.120-138, (1996).
14. G. Takeuti and M. Yasumoto, Forcing on Bounded Arithmetic II. Journal of Symbolic Logic, vol.63(3), pp.860-868, (1998).

Permission and Obligation in Ceteris Paribus

Huimin Dong

Department of Philosophy, Sun Yet-sen University, Zhuhai, China
E-mail: huimin.dong@xixilogic.org

1. Motivation

This paper is motivated to formalize permission and obligation as sufficient and necessary conditions for being normatively fine,[1,6] governed by the concept called *ceteris paribus*. What we propose aims at providing a general theory to illustrate the foundations behind some deontic paradoxes in natural language, like the Lewis problem, the Ross paradox, and the gentle murder problem.[15,16] Furthermore, it also intends to provide a new formal tool to solve the so-called the equilibrium selection problem in game theory.[13]

Two well-known families of deontic logics, deontic action logics (DAL)[1,5] and deontic preference logics (DPL),[7,20] also follow the same principle for permission and obligation, but evaluate what is ought/permitted to do by different conceptual criterions, the Right or the Good. These two approaches solve the infamous deontic paradoxes, like the Lewis problem,[15] the Ross paradox, and the gentle murder paradox,[16] by developing different theoretical methodologies. The methods DAL usually emphasizes that different complicate algebraic interpretations on actions can combine with the Right by ideality,[1,5] but which cost a pay on the intuition of actions and result in very weak logics. On the other hand, DPL usually adopts the similarity-based preference to define obligation as what is the best given everything else being equal.[7,19,20] Such a similarity-based method is natural and offers rich enough logics. However, they are hard to get rid of the Ross paradox and the Lewis problem.

This paper proposes a plausible middle ground in between, which adopts the Right as the conceptual standard, and takes the natural intuition of similarity for actions. Here we understand *ceteris paribus* in terms of similarity. We follow two senses of it proposed by Schurz (see[17]), the static one and the dynamic one, in order to describe the natures of permission and obligation. The two meanings of *ceteris paribus* are later axiomatized in logics, which retain rich principles of permission and obligation as sufficient and necessary conditions for being normatively right, and capture the above two characters of *ceteris paribus*. We leave the detailed proofs to the full paper.

2. Formal Theory

Permission is what does stay to be normally fine, underlying on "everything else are equal." The static *ceteris paribus* in this sense is captured by the concept of likelihood (or similarity), followed after the accounts developed in.[10,14,19] So permission is defined in the style of free choice permission.[18] The case that "if φ then *ceteris paribus* ψ" is interpreted as "in the most likely states satisfying φ, ψ is the case." Obligation is standard.[16] But permission is based on the notion of *ceteris paribus*, which is not a dual of obligation as in standard deontic logic.[11]

Definition 2.1 (Language). *The set $\mathcal{L}$ of well-formed formulas of norms are defined as follows:*

$$\varphi := p \mid \varphi \wedge \varphi \mid \neg\varphi \mid P\varphi \mid O\varphi \mid \varphi \trianglelefteq \psi$$

where $p \in Prop$ is an element of the (countable) set of atomic propositions.

We read $P\varphi$ as "it is permitted that φ" and $O\varphi$ that "it is obligated that φ." The formula $\varphi \trianglelefteq \psi$ is interpreted as "ψ is at least more likely than φ" (see[10]). The formula $\varphi \trianglelefteq \psi \wedge \psi \trianglelefteq \varphi$ is denoted as $\varphi \bowtie \psi$, which indicated that φ and ψ is equally likely. Following,[19] we define $A\varphi := \neg\varphi \trianglelefteq \bot$, and so its dual $E\varphi := \neg A\neg\varphi$.

Definition 2.2 (Deontic Models). *A deontic model M is a tuple $\langle W, R, \leq, V \rangle$ where:*

- W *is a non-empty finite set of states;*
- $R \subseteq W \times W$ *is serial;*
- $\leq$ *is a similarity relation* * *on* $W \times W$ *s.t.* $\leq$ *is reflexive, transitive, and connected;*
- $V : Prop \to \wp(W)$ *is a valuation function.*

Rwu means that from state w state u is normatively fine. We read $u \leq s$ as "u is at least as similar or likely to s".[9] We can see that a similarity relation $\leq$ is a partial well-order. With the finite domain W we can make sure that a maximal element exists in it (see[10]). We define the strict order $<$ as: $u < v$ iff $u \leq v$ and $v \not\leq u$. We then define maximality by using the similarity relation: $\max_{\leq}(X) = \{v \in X \mid \forall u \in X \text{ s.t. } u \leq v\}$. Moreover, we denote $S[w] = \{u \mid Swu\}$ where $S \in \{R, \leq\}$, and $||\varphi|| = \{w \mid M, w \models \varphi\}$. The truth conditions for permission and obligation are defined as follows:

$$M, w \models P\varphi \text{ iff } \max_{\leq}(||\varphi||) \subseteq R[w]$$
$$M, w \models O\varphi \text{ iff } R[w] \subseteq ||\varphi||$$

The similarity modality $\trianglelefteq$ is the $\leq_{\forall\exists}$ modality well discussed in[19] and further developed in:[9]

$$M, w \models \varphi \trianglelefteq \psi \text{ iff } \forall u \in W \exists v \in W \text{ s.t. } (M, u \models \varphi \Rightarrow M, v \models \psi \text{ \& } u \leq v)$$

This implies that A is a universal modality and E is an existential modality:

$$M, w \models A\varphi \text{ iff } \forall u \in W \text{ s.t. } M, u \models \varphi$$
$$M, w \models E\varphi \text{ iff } \exists u \in W \text{ s.t. } M, u \models \varphi$$

We then introduce $\Box(\varphi/\psi)$, read as "if φ then, *ceteris paribus*, ψ," into the language. The formula $\Box(\top/\varphi)$ can be simplified as $\Box\varphi$, and its dual $\Diamond\varphi = \neg\Box\neg\varphi$. The abbreviation $\Box\varphi$ can be read as "it is *ceteris paribus* that φ. The formula $\Box(\varphi/\psi) \wedge \Box(\psi/\varphi)$ is denoted as $\Box(\varphi \mid \psi)$. The modality $\Box$ is interpreted by maximality as follows:

$$M, w \models \Box(\varphi/\psi) \text{ iff } \max_{\leq}(||\varphi||) \subseteq ||\psi||.$$

---

*v is at least as similar as u is when $u \leq v$, and v is strictly more similar than u when $u < v$.[9]

This condition can be used to by similarity:[4,21]

$$M, w \models \Box(\varphi/\psi) \text{ iff } \forall x \; [M, x \models \varphi$$
$$\Rightarrow \exists u \geq x \; (M, u \models \varphi \; \& \; \forall s \geq u \; (M, s \models \varphi \rightarrow \psi))]$$

Obvious it results a reduction of the static *ceteris paribus* by similarity relation:

$$\Box(\varphi/\psi) \leftrightarrow A[\varphi \rightarrow \neg(\varphi \trianglelefteq (\varphi \wedge \neg\psi))]$$

The second sense of *ceteris paribus*, "everything else being right," is that it cannot be *changed* by other things interfere it. After considering all interferences, namely exceptions, the meaning of a *ceteris paribus* sentence is settled down. Now we develop a dynamic logic for the *ceteris paribus* sentences. First we need to have a method to clarify the scope of *exception* in a *ceteris paribus* sentence. We first define $U(\varphi)$ as the set of all atomic propositions occurring in φ as follows:

$$U(p) = \{p\}$$
$$U(\varphi \wedge \psi) = U(\varphi) \cup U(\psi)$$
$$U(\neg\varphi) = U(\varphi)$$
$$U(P\varphi) = U(\varphi)$$
$$U(O\varphi) = U(\varphi)$$
$$U(\Box(\varphi/\psi)) = U(\varphi) \cup U(\psi)$$

Given a finite set Γ of formulae, we define $U(\Gamma)$ as the set of all atomic propositions occurring in Γ, i.e. $U(\Gamma) = \bigcup_{\varphi \in \Gamma} U(\varphi)$. After we can define $C(\varphi)$ as the scale about the disturbing factors regarding φ, represented by all atomic propositions occurring in φ as follows: $C(\varphi) = \{\{\pm p \mid p \in U(\varphi)\} \mid \text{either } \pm p = p \text{ or } \pm p = \neg p\}$. We then define $C(\Gamma)$ the scope of Γ as follows: $C(\Gamma) = \{\{\pm p \mid p \in U(\Gamma)\} \mid \text{either } \pm p = p \text{ or } \pm p = \neg p\}$. Observe that $C(\Gamma)$ exhausts all possible exceptions regarding Γ, while each exception in $C(\Gamma)$ is mutually exclusive to each other. After clarifying the scope of exceptions or interferences, we introduce the exclusive models to illustrate the update of exception.

Definition 2.3 (Exclusive Models). *An exclusive model $\mathcal{C}^{\Gamma}$ is a tuple $\langle C, \preceq \rangle$ defined as follows:*

- $C = C(\Gamma)$;
- $\preceq \subseteq C \times C$ *is reflexive, transitive, and connected.*

We use $c \sim d$ to denote $c \preceq d$ and $d \preceq c$. Because Γ is finite, $C(\Gamma)$ should be finite as well. This leads to that $\preceq$ is conversely well-founded [†]. As the similarity can affect normality and the normative notions in the update, we understand $c \preceq d$ as "The exception c is more *normatively salient* than the exception d." Notice that $\prec = \preceq - \sim$.

Definition 2.4 (Updated Models). *Given $M = \langle W, R, \leq, V \rangle$ and $\mathcal{C}^{\Gamma} = \langle C, \preceq \rangle$. We define the updated model $M \otimes \mathcal{C}^{\Gamma} = \langle W^*, R^*, \leq^*, V^* \rangle$ as follows:*

- $W^* = \{(u, c) \mid M, u \models c \text{ where } c \in C\}$; [‡]
- $(u, c) \leq^* (v, d)$ *iff either $c \prec d$ or ($c \sim d$ but $u \leq v$);*
- $(u, c)R^*(v, d)$ *iff uRv and $c \preceq d$;* [§]
- $(u, c) \in V^*(p)$ *iff $u \in V(p)$.*

The truth condition of Γ-scope updated is denoted as $|| \cdot ||^{\Gamma}$. Notice that if there are $c, d \in C$ s.t. $M, w \models c$ and $M, w \models d$, then it must be the case that c and d are the same exception in C. We call this phenomenon *exception preserving*. After the update, the updated similarity $\leq^*$ in the new model is conversely well-founded. This observation indicates that permission and normality are still well-defined in the updated model. The updated order is generated by the so-called *lexicographic upgrade* methodology,[20] which here is presented by together a similarity over exceptions with a similarity over states. Moreover, because $\preceq$ in an exclusive model is reflexive, it implies that it is also serial. It then results that R^* is serial in the updated model. Note that what was normatively fine can be cancelled after update, if it was not normatively salient enough.

[†]It means that $\preceq$ is not an infinite right branching order.[8]

[‡]Here we simplify $M, w \models \bigwedge_{\pm p \in c} \pm p$ as $M, w \models c$.

[§]A similar suggestion please refer to.[23]

The dynamic sentence $\langle\Gamma\rangle\varphi$ is added up into the language. The dual of $\langle\Gamma\rangle\varphi$ is $[\Gamma]\varphi$, namely $\neg\langle\Gamma\rangle\neg\varphi$. We read $\langle\Gamma\rangle\varphi$ as "It is the case that φ, *Provisos* Γ." ¶ Its truth condition is defined as:

$$M, w \models \langle\Gamma\rangle\varphi \text{ iff } \exists(w, c) \in W^* \text{ s.t. } M \otimes C^\Gamma, (w, c) \models \varphi.$$

Observe that $\langle\Gamma\rangle$ has the same truth condition as $[\Gamma]$ does, by the so-called exception-preserving regarding Γ. As that $C(\Gamma)$ exhausts all possible consistent exceptions regarding Γ, there must be a $c \in C(\Gamma)$ s.t. c is satisfied at w. Together with the exception preserving property, it is not hard to see that $[\Gamma]$ and $\langle\Gamma\rangle$ have the same truth conditions. In particular, we have the following equivalence: $M \otimes C^\Gamma, (w, c) \models \varphi$ iff $M, w \models c \wedge \langle\Gamma\rangle\varphi$.

The axiomatization of permission, obligation, and *ceteris paribus* is structured into two parts. The static part involves axioms and rules in a standard manner. The binary modality $\trianglelefteq$ follows the axioms and rules suggested in,[10] while $\square$ follows those in.[4] The O-modality is the D-modality in modal logic,[3] but the P-modality is not the dual of O anymore. Rather, P-modality satisfies the axioms presented in Table 1.

Theorem 2.1. *The system in Table 1 is (weakly) sound and complete.*

Table 1.

- Tautologies
- The binary modality $\trianglelefteq$ satisfies the axioms and rules suggested in[10]
- The binary modality $\square$ satisfies the axioms and rules suggested in[4]
- O is a D-modality
- OiC: $O\varphi \to E\varphi$
- PtF: $P\bot$
- RFCP: $P\varphi \wedge P\psi \to P(\varphi \vee \psi)$
- FCP: $P\varphi \wedge \square(\psi/\varphi) \to P\psi$
- OWP: $O\varphi \wedge P\psi \to \square(\psi/\varphi)$

¶*Provisos* means "providing some disturbing factors are absent."

In our paper the completeness of the static logic (in Table 1) is proven by finite canonical model followed after.[22] The dynamic part (in Table 2) follows the methodology of reduction in[2,20] for dynamic *ceteris paribus*. The dynamic operator illustrates the complex update apparatus in the syntax level, bridging the notion of exception with the static *ceteris paribus*.

Theorem 2.2. *The system in Table 2 is (weakly) sound and complete.*

Table 2.

- $[\Gamma]p \leftrightarrow \bigwedge_{c \in C}(c \to p)$

- $[\Gamma]\varphi \wedge \psi \leftrightarrow [\Gamma]\varphi \wedge [\Gamma]\psi$

- $[\Gamma]\neg\varphi \leftrightarrow \bigwedge_{c \in C}(c \to \neg[\Gamma]\varphi)$

- $[\Gamma]O\varphi \leftrightarrow \bigwedge_{c \in C}(c \to \bigwedge_{d \succeq c} O(d \wedge \langle\Gamma\rangle\varphi))$

- $[\Gamma]P\varphi \leftrightarrow \bigwedge_{c \in C}\{c \to \bigwedge_{d \in C}[(A \bigwedge_{e \succ d}\Gamma^e_\varphi \to$

  $P \bigvee_{e \sim d} \neg\Gamma^e_\varphi) \wedge \bigwedge_{d \not\succeq c} \Box(\bigvee_{e \sim d} \neg\Gamma^e_\varphi / E \bigwedge_{e \succ d}\Gamma^e_\varphi)]\}$

- $[\Gamma](\varphi \trianglelefteq \psi) \leftrightarrow \bigwedge_{c \in C}\{c \to \bigwedge_{d \in C}[A(\bigvee_{e \sim d} \neg\Gamma^e_\varphi \to E \bigvee_{e \succ d} \neg\Gamma^e_\psi)$

  $\vee (\bigvee_{e \sim d} \neg\Gamma^e_\varphi) \trianglelefteq (\bigvee_{e \sim d} \neg\Gamma^e_\psi)]\}$

  where $\Gamma^c_\varphi := c \to [\Gamma]\neg\varphi$

3. Conclusion

This paper has argued that permission and obligation should be viewed as the sufficient and necessary conditions for the Right, but should be governed under the principle *ceteris paribus*. This view point turns to a general formal theory of a (weakly) sound and complete dynamic logic for permission and obligation, which can answer the Lewis problem, the gentle murder puzzle, the Ross paradox, and the equilibrium selection problem. This theory is theoretically rich, because it includes a great number of interactions between permission, obligation, and *ceteris paribus*. It is also expressive enough such that a fragment of DPL is one of its special cases, specified that the bestness is a particular case of the rightness. All details are left in the full paper.

In the future we want to discuss the issue of whether the similarity relation is connected. Intuitively a similarity is more common to be non-comparable.[12] Connectedness captures an important kind of norm called overall norms, which satisfy the standard consistency principle (axiom D). Non-connectedness captures the *prima facie* norms more properly. A natural question rises up: What about the consistency principle for the non-connected *prima facie* norms? We leave it open for the future research.

Acknowledgements

The author is supported by the China Postdoctoral Science Foundation funded project [No. 2018M632494], the PIOTR research project [No. RO 4548/4-1], the MOE Project of Key Research Institute of Humanities and Social Sciences in Universities [No. 17JJD720008], the National Social Science Fund of China [No. 17ZDA026], and the Fundamental Research Funds for the Central Universities of China. The author thanks Johan van Benthem, Fan Jie, Olivier Roy, Patrick Blackburn for the insightful comments and inspirations. The author also thanks for the reviewers of AWPL 2018 and NCML 2018 and the participants of NCL 2018, ISLU 2018, and the PIOTR workshop 2018 for the helpful comments.

References

1. A. J. Anglberger, N. Gratzl, and O. Roy. Obligation, free choice, and the logic of weakest permissions. *The Review of Symbolic Logic*, 8:807–827, December 2015.
2. A. Baltag and S. Smets. A qualitative theory of dynamic interactive belief revision. *Logic and the foundations of game and decision theory (LOFT 7)*, 3:9–58, 2008.
3. P. Blackburn, M. De Rijke, and Y. Venema. *Modal Logic*, volume 53. Cambridge University Press, 2002.
4. J. P. Burgess. Quick completeness proofs for some logics of conditionals. *Notre Dame Journal of Formal Logic*, 22(1):76–84, 1981.
5. F. Dignum, J.-J. C. Meyer, and R. J. Wieringa. Free choice and contextually permitted actions. *Studia Logica*, 57(1):193–220, 1996.

6. H. Dong and O. Roy. Three deontic logics for rational agency in games. *Studies in Logic*, 8(4):7–31, 2015.

7. H. Dong and O. Roy. Dynamic logic of power and immunity. In *International Workshop on Logic, Rationality and Interaction*, pages 123–136. Springer, 2017.

8. P. Girard and H. Rott. Belief revision and dynamic logic. In *Johan van Benthem on logic and information dynamics*, pages 203–233. Springer, 2014.

9. P. Girard and M. A. Triplett. Prioritised ceteris paribus logic for counterfactual reasoning. *Synthese*, pages 1–23, 2017.

10. J. Y. Halpern. Defining relative likelihood in partially-ordered preferential structures. *Journal of Artificial Intelligence Research*, 7:1–24, 1997.

11. S. O. Hansson. The varieties of permissions. In D. Gabbay, J. Horty, X. Parent, R. van der Meyden, and L. van der Torre, editors, *Handbook of Deontic Logic and Normative Systems*, volume 1. College Publication, 2013.

12. S. O. Hansson. Deontic diversity. In *International Conference on Deontic Logic in Computer Science*, pages 5–18. Springer, 2014.

13. J. C. Harsanyi and R. Selten. *A general theory of equilibrium selection in games*. The MIT Press, 1988.

14. D. Lewis. *Counterfactuals*. Blackwell, 1973.

15. D. Lewis. A problem about permission. In *Essays in honour of Jaakko Hintikka*, pages 163–175. Springer, 1979.

16. P. McNamara. Deontic logic. In E. N. Zalta, editor, *The Stanford Encyclopedia of Philosophy*. Metaphysics Research Lab, Stanford University, winter 2014 edition, 2014.

17. G. Schurz. Ceteris paribus laws: Classification and deconstruction. *Erkenntnis*, 57:351–371, 2002.

18. J. van Benthem. Minimal deontic logics. *Bulletin of the Section of Logic*, 8(1):36–42, 1979.

19. J. van Benthem, P. Girard, and O. Roy. Everything else being equal: A modal logic for ceteris paribus preferences. *Journal of philosophical logic*, 38(1):83–125, 2009.

20. J. van Benthem, D. Grossi, and F. Liu. Priority structures in deontic logic. *Theoria*, 80(2):116–152, 2014.

21. J. van Benthem, S. van Otterloo, and O. Roy. Preference logic, conditionals and solution concepts in games. 2005.

22. F. Van De Putte. "that will do": Logics of deontic necessity and sufficiency. *Erkenntnis*, 82(3):473–511, 2017.

23. S. van Wijk. Coalitions in epistemic planning. Master's thesis, Universiteit van Amsterdam, 2015.

Reverse Mathematics of Separation Theorems in Lattice Theory

Junren Ru and Guohua Wu[*]

Division of Mathematical Sciences,
School of Physical and Mathematical Sciences,
Nanyang Technological University,
21 Nanyang Link, Singapore 637371
[*] E-mail: guohua@ntu.edu.sg

Reverse mathematics is a program of determining which axioms are required to prove theorems of mathematics. In this paper, we study reverse mathematics of two well-known theorems in countable lattices. We will first give a brief survey of existing work of doing reverse mathematics in partially ordered sets, and then prove that the existence of $\mathcal{J}(L)$, the set of irreducible elements, of a countable lattice L, is equivalent to ACA_0. We will also prove that **DPI**, prime ideal separation theorem, is provable in WKL$_0$.

1. Introduction

In the proof of Birkhoff's Representation Theorem, a most important component is $\mathcal{J}(L)$, the set of join-irreducible elements, which was used in the definition of the mapping of elements in L to subsets of $\mathcal{J}(L)$. In this paper, we consider the difficulty of finding the existence of $\mathcal{J}(L)$. If L is finite, then $\mathcal{J}(L)$ is finite, and we can find $\mathcal{J}(L)$ by a simple observation. It becomes harder when L is infinite. We prove in this paper that for a countable lattice L, we need arithmetic comprehension to find $\mathcal{J}(L)$: the existence of $\mathcal{J}(L)$ is equivalent to ACA_0 over RCA_0, a project of reverse mathematics of lattice theory.

Reverse mathematics was proposed by Friedman in his PhD thesis,[6] and further developed by Friedman, Simpson, and several other researchers from recursion theory. This program has attracted researchers from both recursion theory and proof theory, with the target of finding the weakest possible "set-theoretical" axiom systems needed to prove theorems in "ordinary" mathematics. Here, ordinary mathematics, such as calculus, differential equations, real analysis, countable algebra, mathematical logic, and so on, means that it is independent of the introduction of abstract set-theoretic concepts. That is, in ordinary mathematics, sets are restricted to be countable-based (for example, the real line is uncountable, but it is a separable metric space), and the reason for this restriction is that set existence axioms needed for uncountable, set-theoretic mathematics are likely to be much stronger than those which are needed for ordinary mathematics. Two good references for reverse mathematics are Simpson's book[19] on many subjects, and Hirschfeldt's monograph[12] on combinatorics. Friedman Simpson and Smith's paper FSS provides a deep investigation of reverse mathematics in countable algebra.

In this paper, "ordinary" refers to "partially ordered sets" and "lattice theory", and the "set-theoretical" axiom systems are weak subsystems of second-order arithmetic. We will use the axiom system RCA_0 (for recursive comprehension with Σ_1^0-induction) as a base system, which is commonly used in practice in this area, and ACA_0 (arithmetic comprehension with Σ_1^0-induction), in which we will show that the existence of certain objects in our concern can be proved.

The following definitions of second-order arithmetic are from Simpson's book.[19]

Definition 1.1. The axioms of second-order arithmetic consist of the universal closures of the following L_2-formulas (L_2 is the language of second-order arithmetic):
(1) Basic axioms:

$$n + 1 \neq 0,$$
$$m + 1 = n + 1 \rightarrow m = n,$$
$$m + 0 = m,$$
$$m + (n + 1) = (m + n) + 1,$$
$$m \cdot 0 = 0,$$
$$m \cdot (n + 1) = (m \cdot n) + m,$$
$$\neg m < 0,$$
$$m < n + 1 \longleftrightarrow (m < n \ \vee \ m = n).$$

(2) Induction axiom:

$$(0 \in X \land \forall n \, (n \in X \to n + 1 \in X)) \to \forall n \, (n \in X).$$

(3) Comprehension scheme:

$$\exists X \, \forall n \, (n \in X \leftrightarrow \varphi(n)),$$

where $\varphi(n)$ is any formula of L_2 in which X does not occur freely. (4) By *second-order arithmetic*, denoted as Z_2, we mean the formal system in the language L_2 which are deducible from those axioms by means of the usual logical axioms and rules of inference.

By *a subsystem* of Z_2, we mean a formal system in the language L_2, each of whose axioms is a theorem of Z_2. In this paper, we will need three subsystems: RCA_0, WKL_0 and ACA_0.

Definition 1.2.

(1) The system RCA_0 consists of the basic axioms, together with schemes of Σ_1^0 induction and Δ_1^0 comprehension, i.e.,

$$\forall n (\phi(n) \leftrightarrow \psi(n)) \to \exists X \forall n (n \in X \leftrightarrow \phi(n)),$$

where $\phi(n)$ is any Σ_1^0 formula, $\psi(n)$ is any Π_1^0 formula.
(2) The system WKL_0 consists of RCA_0 and weak König's lemma: Every infinite subtree of $2^{<\omega}$ has an infinite path.
(3) The system ACA_0 consists of the basic axioms and the induction axiom, together with arithmetic comprehension, i.e., $\varphi(n)$ is an arithmetic formula in the comprehension scheme.

Proposition 1.1. *RCA_0 proves bounded Σ_1^0 comprehension:*

$$\forall n \exists X \forall i (i \in X \iff i < n \ \& \ \phi(i)),$$

where $\phi(i)$ is any Σ_1^0 formula in which X does not occur freely.

The following two theorems provide useful equivalences of ACA_0 and WKL_0, respectively. When we try to show the reversal parts, we often show that one of the equivalences can be proved.

Proposition 1.2. *The following are equivalent over RCA_0:*

(1) WKL$_0$: Every infinite subtree of $2^{<\omega}$ has an infinite path;

(2) If $f, g : \mathbb{N} \to \mathbb{N}$ are one-to-one functions with $\forall m \forall n f(m) \neq g(n)$, then there exists a set $X \subseteq \mathbb{N}$ such that

$$\forall n(f(n) \in X \ \& \ g(n) \notin X).$$

Proposition 1.3. *The following are equivalent over RCA$_0$:*

(1) ACA$_0$;

(2) For every one-to-one function $f : \mathbb{N} \to \mathbb{N}$, there exists a set $X \subseteq \mathbb{N}$ such that

$$\forall n(n \in X \iff \exists m(f(m) = n));$$

(3) König lemma: Every infinite, finitely branching tree $T \subseteq \omega^{<\omega}$ has an infinite path.

Below is a quick review of reverse mathematics of theorems in partially ordered sets (posets for short). For a poset $(P, \leq)$, we use $\downarrow A$ to denote the set $\{x \in P : \exists a \in A \ (x \preceq a)\}$, the *downward closure* of A, and A is called a lower set if $A = \downarrow A$. Note that $\downarrow A$ is the smallest lower set containing A. Upper sets are defined dually. In his thesis, Mummert proved that the existence of downward closure requires ACA$_0$.

Theorem 1.1. *(Mummert[16]) The following are equivalent over RCA$_0$:*

(1) ACA$_0$;

(2) Every subset of a countable poset has a downward closure.

Lempp and Mummert proved in their paper[13] that the existence of maximal ideals of countable posets is equivalent to ACA$_0$, over RCA$_0$, of course. In his thesis, Hummert also considered the logical strength of the the extension of ideals to maximal ideals, and proved that it is much more complicated than the existence of maximal ideals.

People also have considered reverse mathematics of other well-known theorems, such as Szpilrajn's theorem on linear extensions of partial orders, Dushnik-Miller's theorem of nontrivial self-embeddings, Dilworth's theorem of decompositions, etc.

2. $\mathcal{J}(L)$ and ACA_0

In this paper, we will consider reverse mathematics of theorems in lattice theory. For notations and concepts for lattices, please refer to Davey and Priestley's book.[4] The lattices we consider here are bounded.

This program of doing reverse mathematics in lattice theory was recently initiated by Brodhead, et al. in paper,[2] and Sato and Yamazaki in.[18] In,[2] Brodhead, et al. proved that the existence of the set of compact elements of a countable lattice, and also the Grätzer-Schmit theorem, i.e., the congruence lattice representation theorem, are equivalent to Π_1^1-CA_0, and in,[18] Sato and Yamazaki pointed out that the Knaster-Tarski's fixed point theorem is provable in RCA_0.

Unlike the situation for posets, for lattices, we have meet and join as operations, and these two operations lower down the logical strength of many statements. For example, in posets, finding meet and join needs Σ_1-comprehension, as they are the infimum and the supremum of infinite sets, respectively. For lattices, we have meet and join directly, and this implies that the existence of maximal ideals of countable lattices can be proved in RCA_0. Our purpose for this program is to allocate theorems in lattice theory in various axiom systems systems, and we will show in the section that the existence of $\mathcal{J}(L)$, the set of join-irreducible elements of L, equivalent to ACA_0 (over RCA_0).

$\mathcal{J}(L)$ is a crucial component in the proof of Birkhoff's Representation Theorem of finite distributive lattices, in the definition of mapping elements in L to subsets of $\mathcal{J}(L)$.

If L is finite, then $\mathcal{J}(L)$ is finite, and we can find $\mathcal{J}(L)$ by a simple observation. It becomes harder when L is infinite.

Theorem 2.1. *The following are equivalent over* RCA_0:

(1) ACA_0;
(2) *For every countable lattice* L, $\mathcal{J}(L)$ *exists.*

Proof. We first assume ACA_0. Then

$$x \in \mathcal{J}(L) \iff x \neq 0 \text{ and } \forall a, b \in L(x = a \vee b \Rightarrow x = a \text{ or } x = b),$$

and $\mathcal{J}(L)$ exists, by Π_1^0-comprehension.

We now prove the other direction. We assume that $\mathcal{J}(L)$ exists for every countable lattice L, and prove that the range of any one-to-one function $f : \mathbb{N} \to \mathbb{N}$ exists.

Consider $L = \{0, 1\} \cup \{a_k^i, b_i : k, i \in \mathbb{N}\}$ with the order $\preceq$ such that

- $a_k^i \preceq b_j \iff i = j$ and $(\exists m \leq k)\ (f(m) = i)$;
- $a_m^i \preceq a_n^j \iff i = j$ and $m = n$;
- $b_i \preceq b_j \iff i = j$.

L above is a lattice, and can be defined over RCA_0:

$$a_k^i \vee b_j = \begin{cases} b_j \text{ if } i = j \text{ and } \exists m \leq k\ (f(m) = i), \\ 1 \text{ otherwise.} \end{cases}$$

$$a_k^i \wedge b_j = \begin{cases} a_k^i \text{ if } i = j \text{ and } \exists m \leq k(f(m) = i, \\ 0 \text{ otherwise.} \end{cases}$$

$$a_m^i \vee a_n^j = \begin{cases} a_m^i \text{ if } i = j, m = n, \\ b_i \text{ if } i = j, m \neq n \\ \qquad \text{and there exist } p \leq m, q \leq n \text{ with } f(p) = f(q) = i, \\ 1 \text{ otherwise.} \end{cases}$$

$$a_m^i \wedge a_n^j = \begin{cases} a_m^i \text{ if } i = j, m = n, \\ 0 \text{ otherwise.} \end{cases}$$

$$b_i \vee b_j = \begin{cases} b_i \text{ if } i = j, \\ 1 \text{ otherwise.} \end{cases}$$

$$b_i \wedge b_j = \begin{cases} b_i \text{ if } i = j, \\ 0 \text{ otherwise.} \end{cases}$$

It is easy to check that

$$i \in rng(f) \iff b_i \text{ is not join-irreducible.}$$

By assumption, $\mathcal{J}(L)$ exists. Now let $X = \{i \in \mathbb{N} : b_i \notin \mathcal{J}(L)\}$ and X exists by Σ_0^0-comprehension. Hence $\forall n (n \in X \iff \exists m(f(m) = n))$, and $rng(f)$ exists, ACA_0 holds. $\qquad\square$

3. Prove DPI in WKL$_0$

DPI says that for a distributive lattice $(L, \preceq, \vee, \wedge, 0, 1)$, if I is an ideal, and F is a filter of L with $I \cap F = \emptyset$, then there exists a prime ideal P containing I but disjoint from F. **DPI** is a separation property, and we prove in this section that **DPI** for countable lattices can be proved in WKL$_0$.

Theorem 3.1. *The following statement can be proved in WKL$_0$: For a countable distributive lattice $(L, \preceq, \vee, \wedge, 0, 1)$, I an ideal, F a filter of L with $I \cap F = \emptyset$, there exists a prime ideal P containing I but disjoint from F.*

Proof. Let $\{a_0, a_1, a_2, \cdots\}$ be an enumeration of L. We assume that $a_0 = 0, a_1 = 1$. Let $T \subseteq 2^{<\omega}$ be the set of all strings $\sigma \in 2^{<\omega}$ such that:

- $a_i \in I$ implies $\sigma(i) = 1$;
- $a_i \in F$ implies $\sigma(i) = 0$;
- For all $i, j, k < lh(\sigma)$,

  - (1) if $\sigma(i) = \sigma(j) = 1$ and $a_i \vee a_j = a_k$, then $\sigma(k) = 1$;
  - (2) if $\sigma(i) = 1$ and $a_j \preceq a_i$, then $\sigma(j) = 1$;
  - (3) if $\sigma(i) = \sigma(j) = 0$ and $a_i \wedge a_j = a_k$, then $\sigma(k) = 0$.

Clearly, T is a tree and T exists by Σ_0^0-comprehension. In order to show that T is infinite, we need the following binary subtree $S \subseteq 2^{<\omega}$. We define a sequence of finite sets $X_s \subseteq L$ for each $s \in 2^{<\omega}$, beginning with $X_{\langle\rangle} = \{0\}$. Suppose that X_s has been defined. Let

$$lh(s) = 4 \cdot \langle\langle i, j\rangle, m\rangle + k, \ 0 \leq k < 4,$$

where $\langle \cdot, \cdot \rangle$ denotes the pairing function.

Case 1: $k = 0$. If $a_i \wedge a_j \in X_s$, then put $X_{s^\frown 0} = X_s \cup \{a_i\}$, $X_{s^\frown 1} = X_s \cup \{a_j\}$; otherwise put $X_{s^\frown 0} = X_s$, $X_{s^\frown 1} = \emptyset$.

Case 2: $k = 1$. Put $X_{s^\frown 0} = \emptyset$. If $a_i \in X_s$, $a_j \in X_s$, then put $X_{s^\frown 1} = X_s \cup \{a_i \vee a_j\}$; otherwise, put $X_{s^\frown 1} = X_s$.

Case 3: $k = 2$. Put $X_{s^\frown 0} = \emptyset$. If $a_i \in X_s$, $a_j \preceq a_i$, then put $X_{s^\frown 1} = X_s \cup \{a_j\}$; otherwise, put $X_{s^\frown 1} = X_s$.

Case 4: $k = 3$. Put $X_{s^\frown 0} = \emptyset$. If $X_s \cap F \neq \emptyset$, i.e., $\exists a \in X_s$ $(a \in F)$, which is a Σ_0^0- sentence (as X_s is finite), then put $X_{s^\frown 1} = \emptyset$; otherwise, put $X_{s^\frown 1} = X_s$.

Let $S = \{s \in 2^{<\omega} : X_s \neq \emptyset\}$. Clearly, S is a tree and S exists by Σ_0^0-comprehension. We claim that for each $n \in \mathbb{N}$, there exists an $s \in S$ with $lh(s) = n$ such that $Cl(X_s) \cap F = \emptyset$. Since X_s is finite and F is a filter, the claim can be written as: For $n \in \mathbb{N}$, $\varphi(n)$, where

$$\varphi(n) := \exists s \in S \ (lh(s) = n \text{ and } \bigvee X_s \notin F).$$

Here, we prove $\varphi(n)$ by Σ_0^0-induction on n. $\varphi(0)$ is trivial. Suppose that $\varphi(n)$ is true for n. If $n \equiv 1, 2,$ or $3 \bmod 4$, then $\varphi(n + 1)$ follows immediately. If $n \equiv 0 \bmod 4$, then there exists $s \in S$ such that $lh(s) = n$ and $\bigvee X_s \notin F$. We only need to consider the case $a_i \wedge a_j \in X_s$, i.e., $X_{s^\frown 0} = X_s \cup \{a_i\}$, $X_{s^\frown 1} = X_s \cup \{a_j\}$.

Claim: $\bigvee X_{s^\frown 0} \notin F$ or $\bigvee X_{s^\frown 1} \notin F$.

Otherwise, we have $(\bigvee X_s) \vee a_i = \bigvee X_{s^\frown 0} \in F$ and $(\bigvee X_s) \vee a_j = \bigvee X_{s^\frown 1} \in F$. By $a_i \wedge a_j \in X_s$ and F is filter, $\bigvee X_s = (\bigvee X_s) \vee (a_i \wedge a_j) \in F$ (by distributivity), a contradiction.

Hence $\bigvee X_{s^\frown 0} \notin F$ or $\bigvee X_{s^\frown 1} \notin F$, and so $\varphi(n + 1)$ is true.

"$\forall n \in \mathbb{N} \ \varphi(n)$" implies that S is infinite. By WKL$_0$, S has an infinite path, g say. Now, coming back to T, we will show that T is infinite.

For $m \in \mathbb{N}$, let $Y_m = \{i < m : \exists n \ (a_i \in X_{g\restriction n})\}$. Y_m exists by bounded Σ_1^0-comprehension. Define $\gamma \in 2^{<\omega}$ such that $lh(\gamma) = m$ and for all $i < m$,

$$\gamma(i) = \begin{cases} 1 \text{ if } i \in Y_m, \\ 0 \text{ otherwise.} \end{cases}$$

Then $\gamma \in T$. This proves that T is infinite. Hence, by WKL$_0$ again, there exists an infinite path f of T. Let $P = \{a_i : f(i) = 1\}$. P is the desired prime ideal. $\qquad\square$

In **DPI**, by letting $F = \{1\}$, the proper ideal I can be extended to the prime ideal P, which implies the prime ideal extension theorem immediately.

Theorem 3.2. *The following extension theorem can be proved in* WKL_0: *Every proper ideal on a countable distributive lattice extends to a prime ideal.*

The next theorem, a direct corollary of Theorem 3.1, plays a central role in the proofs of Stone's Representation Theorem and Priestley's Representation Theorem.

Theorem 3.3. (WKL_0) *Let* $(L, \preceq, \vee, \wedge, 0, 1)$ *be a countable distributive lattice. Suppose* $a, b \in L$ *and* $a \not\preceq b$, *then there exists a prime ideal* P *such that* $a \notin P$ *and* $b \in P$.

Proof. Let $I = \downarrow b$ and $F = \uparrow a$. I and F exist by Σ_0^0 comprehension. Since $a \not\preceq b$, $I \cap F = \emptyset$. By Theorem 3.1, we can have a prime ideal P containing I avoiding F (hence $a \notin P$ and $b \in P$) and this can be done within WKL_0. $\qquad\square$

4. Beyond: some problems

In this section, we list a couple of related work we have done, and raise some problems in this direction.

- For distributive lattices, we know how to prove **DPI**, i.e., prime ideal separation theorem, in WKL_0, and we ask whether **DPI** and WKL_0 are equivalent over RCA_0.
- In general, lattices many not have prime ideals. In this general case, we have relatively maximal ideals, and **RMI**, i.e., relatively maximal ideal separation theorem. **RMI** can be proved in ACA_0. We ask whether **RMI** and ACA_0 are equivalent over RCA_0.
- Rudin's lemma plays a central role in the study of quasicontinuous lattices in domain theory. In,[14] Li, Ru and Wu proved that Rudin's lemma is equivalent to ACA_0 over RCA_0. It is interesting to study the logical strength of the topological version of Rudin's lemma.

We believe that several theorems in domain theory can be proved in ACA_0, and we ask which theorems in domain theory necessarily need stronger axiom systems to prove.

Acknowledgements

This paper was supported by Singapore Ministry of Education Tier 2 grant MOE2016-T2-1-083 (M4020333); NTU Tier 1 grants RG32/16 (M4011672) and RG111/19 (M4012245).

References

1. R. Bonnet. On the cardinality of the set of initial intervals of a partially ordered set. *Infinite and finite sets: to Paul Erdös on his 60th birthday*, 1: 189–198, 1975.
2. K. Brodhead, M. Khan, B. Kjøs-Hanssen, W. Lampe, P. Nguyen, R. Shore. The strength of the Grätzer-Schmidt theorem. *Archive for Mathematical Logic*, 55(5-6): 687–704, 2016.
3. D. Cenzer and J. B. Remmel. Proof-theoretic strength of the stable marriage theorem and other problems. *Reverse Mathematics 2001*, Cambridge University Press, 67–103, 2017.
4. B. A. Davey, H. A. Priestley. *Introduction to lattices and order*. Cambridge University Press, 2002.
5. R. Downey, S. Lempp. The proof-theoretic strength of the Dushnik-Miller Theorem for countable linear orders. *Recursion Theory and Complexity*, Proceedings of the Kazan '97 Workshop, Kazan, Russia, (Eds: Arslanov, Lempp), 55–58, 1999.
6. H. Friedman. *Subsystems of Set Theory and Analysis*. PhD thesis, Massachusetts Institute of Technology, 1967.
7. H. Friedman. Subsystems of second-order arithmetic with restricted induction. *Journal of Symbolic Logic*, 41(2): 557–560, 1976.
8. H. Friedman, G. Simpson, L. Smith. Countable algebra and set existence axioms. *Annals of Pure and Applied Logic*, 25: 141–181, 1983.
9. G. Gierz, K. H. Hofmann, K. Keimel, J. D. Lawson, M. Mislove, D. Scott. *Continuous Lattices and Domains*. Cambridge University Press, 2003.
10. G. Gierz, J. D. Lawson, A. Stralka. Quasicontinuous posets. *Houston J. Math*, 9(2): 191–208, 1983.
11. R. Heckmann, K. Keimel. Quasicontinuous domains and the smyth powerdomain. *Electronic Notes in Theoretical Computer Science*, 298: 215–232, 2013.
12. D. R. Hirschfeldt. *Slicing the truth: On the computable and reverse mathematics of combinatorial principles*. World Scientific, 2014.
13. S. Lempp, C. Mummert. Filters on computable posets. *Notre Dame Journal of Formal Logic*, 47(4): 479–485, 2006.
14. G. Li, J. Ru and G. Wu. Rudin's lemma and reverse mathematics.

Annals of the Japan Association for Philosophy of Science, 25: 57–66, 2017.

15. G. Markowsky. Chain-complete posets and directed sets with applications. *Algebra Universalis*, 6(1): 53–68, 1976.

16. C. Mummert. *On the reverse mathematics of general topology*. PhD Thesis, The Pennsylvania State University, 2005.

17. H. Priestley. Representation of distributive lattices by means of ordered Stone spaces. *Bulletin of the London Mathematical Society*, 2(2): 186–190, 1970.

18. T. Sato, T. Yamazaki. Reverse mathematics and order theoretic fixed point theorems. *Archive for Mathematical Logic*, 56: 385–396, 2017.

19. S. G. Simpson. *Subsystems of Second Order Arithmetic*. Cambridge University Press, 2009.